PPM

Practical Problems in Mathematics

FOR MASONS

3RD EDITION

FOR MASONS

3RD EDITION

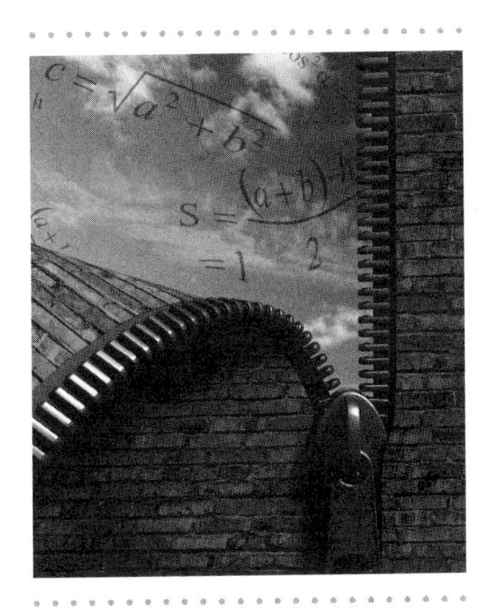

John E. Ball
Robert B. Ham
Donna B. Ham

Australia • Brazil • Japan • Korea • Mexico • Singapore • Spain • United Kingdom • United States

Practical Problems in Mathematics for Masons, Third Edition
John E. Ball, Robert B. Ham, and Donna B. Ham

Vice President, Career and Professional Editorial: Dave Garza

Director of Learning Solutions: Sandy Clark

Associate Acquisitions Editor: Kathryn Hall

Managing Editor: Larry Main

Senior Product Manager: Mary Clyne

Vice President, Marketing: Jennifer Ann Baker

Marketing Director: Deborah Yarnell

Associate Marketing Manager: Erica Glisson

Senior Production Director: Wendy A. Troeger

Production Manager: Mark Bernard

Content Project Manager: David Barnes

Senior Art Director: David Arsenault

Cover image credits: © Kutlayev Dmitry/www.Shutterstock.com, © Vasilius/www.Shutterstock.com

Library of Congress Control Number: 2011942294

ISBN-13: 978-1-133-01474-4

ISBN-10: 1-133-01474-7

Delmar
5 Maxwell Drive
Clifton Park, NY 12065–2919
USA

Cengage Learning products are represented in Canada by Nelson Education, Ltd.

For your lifelong learning solutions, visit **delmar.cengage.com**
Visit our corporate website at **cengage.com.**

Printed in the United States of America
1 2 3 4 5 6 7 16 15 14 13 12

CONTENTS

SECTION 8 FORMULAS TO ALIGN MASONRY WALLS

SECTION 9 MATERIALS ESTIMATION

PREFACE

To succeed in the masonry field, one must have a substantial background in mathematics. The third edition of *Practical Problems in Mathematics for Masons* has been revised to provide beginning students with these needed skills. This clearly organized workbook will help students gain experience and confidence in computing problems common in a wide variety of masonry applications.

Practical Problems in Mathematics for Masons, third edition, starts with basic arithmetic and progresses through area and volume to estimating materials and using the Pythagorean Theorem. The explanations and examples at the beginning of each unit help students build a better understanding of the concepts presented. Most are word problems designed to encourage the student to use logical deduction to arrive at an answer. Many of these word problems are multi-stepped. Problems related to masonry are used throughout the text to help students understand masonry terms and practices.

In addition, students will develop an awareness of masonry terms, building plans, and calculator use as they work through this text. The answers to odd-numbered problems are provided at the end of this text along with a complete appendix of useful references.

DELMAR'S PPM SERIES

This text is one of a series of workbooks designed to offer students practical problem-solving experience in various occupations. The workbooks take a step-by-step approach to mastering basic math skills. Each workbook includes relevant and easily understood problems in a specific vocational field. The workbooks are suitable for any student from junior high through high school and up to the two-year college level. Each text includes a glossary to help students with technical terms. *Practical Problems in Mathematics for Masons* includes an appendix with information on basic operations, English and SI measurements, important formulas, and answers to odd-numbered questions. For more information about this series and a current list of titles, visit www.CengageBrain.com.

SERIES FEATURES

The workbooks in Delmar's PPM series take a step-by-step approach to mastering essential math skills. At the start of each unit, a brief introductory section provides a basic explanation of the concepts necessary to solve the problems in the unit. Examples are presented to help the students review the mathematical principles. The problems in each unit progress from basic examples of the math concepts to more complex examples that require critical thinking. As students progress through each unit, they will become more proficient at solving a wide variety of math problems.

THIS BOOK'S APPROACH

Practical Problems in Mathematics for Masons, third edition, begins with a review of basic operations with whole numbers, fractions, and decimals; progresses through measurements, area and volume calculations, ratio and proportion; and ends with a section on estimation. Topical sections are divided into short units to give instructors maximum flexibility in planning and to help students achieve maximum skill mastery. Instructors may choose to use this book as a stand-alone text or as a supplemental workbook to a theory-based text.

NEW TO THIS EDITION

The third edition of *Practical Problems in Mathematics for Masons* has been updated to include:

- Examples showing current industry applications and practices
- New problems highlighting energy-efficient technologies
- An introduction to using a handheld calculator and new examples.
- Improved illustrations to help student visualize problems and solutions

SUPPLEMENTS

The supplements package for this edition has been revised and expanded to include a new Instructor's Companion Website and Applied Math CourseMate, a new online tool that can help students and teachers build lasting math skills.

Instructor Resources

The Instructor's Companion Website provides the following support for instructors:

- Updated answers to all text problems
- Computerized test banks in ExamView® software
- PowerPoint® presentations
- An image gallery including all text figures

Applied Math CourseMate

Every text in Delmar's PPM series includes Applied Math CourseMate, Cengage Learning's online solution for building strong math skills. Students and instructors alike will benefit from the following CourseMate resources:

- An interactive eBook, with highlighting, note-taking, and search capabilities
- interactive learning tools including:
  - ✓ Quizzes
  - ✓ Flashcards
  - ✓ PowerPoint slides
  - ✓ Skill-building games
- And more!

Instructors will be able to use Applied Math CourseMate to access the Instructor Resources and other classroom management tools. To access CourseMate, visit www.cengagebrain.com. At the CengageBrain.com homepage, search for the ISBN of your title (from the back cover of your book) using the search box at the top of the page. This will take you to the product page where these resources can be found.

ACKNOWLEDGMENTS

The author and publisher acknowledge the contributions of our reviewers for this text:

Paul Geisler
Dakota County Technical College
Rosemount, Minnesota

Don Borchert
Southwest Wisconsin Technical College
Fenimore, Wisconsin

Special thanks go to Delmar's editorial consultant for applied mathematics, John C. Peterson.

ABOUT THE AUTHORS

Robert B. Ham, a 1974 graduate of Virginia Tech, is a retired vocational education teacher. He is a recent safety manager and director of masonry apprenticeship training for an established Virginia commercial masonry contractor. He is the author of *Masonry—Brick and Block Construction*, a 2007 publication of Delmar, Cengage Learning. He is a journeyman mason and masonry contractor. Mr. Ham is the 2002 recipient of The Virginia Association of Trade and Industrial Education's Outstanding Teacher of the Year award.

Donna B. Ham is a 1974 graduate of James Madison University and is a retired high school mathematics teacher. She holds a B.A. degree in mathematics and a M.Ed. degree with a concentration in mathematics. Donna is an adjunct faculty member of Blue Ridge Community College and is a recent Virginia Standards of Learning (SOL) Testing Manager. She has served on the Board of the Augusta County Federal Credit Union and is currently serving as chairman of the Supervisory Committee.

John Ball taught at various levels including middle school, high school, junior college, college, and university. He was involved in building construction courses in the associate degree program at Northern Louisiana University in Monroe, Louisiana. He has been an active member of several professional organizations and has served as a speaker for the Associated General Contractors.

USING THE CALCULATOR

Use of the calculator in math can give you several advantages in your work duties. A properly working calculator in experienced hands is: (1) accurate and (2) a timesaver.

In choosing a calculator several things are considered, such as (1) name brand, (2) cost, (3) scientific vs. basic.

NAME BRAND

As in all tools and equipment the craftsman needs, the name brand is usually there for a reason. The manufacturer usually has the experience, know-how, and resources to provide a reliable quality instrument.

COST

An inexpensive model will do all the work you would normally run into in the field and on the job. Work you would normally do in the field is an original calculation or measurement by yourself, or is in checking calculations given to you in a blueprint, by fellow workers or your foreman.

SCIENTIFIC

All your calculations in the field, and in this book, can be solved using the +, −, ×, and division functions. The possible problem with a scientific calculator is the excess quantity of buttons not normally used that can be accidentally pushed or touched, complicating the process. In some models these buttons take up so much space that the buttons that are actually needed then have to be reduced in size. However, scientific calculators can be used well in the classroom, some have addressed the button size issue, and good quality name brand scientific calculators can cost as little as $10.00 to $15.00.

One difference in types of calculators you may find is that scientific calculators have the "order of operations" automatically programmed (see Unit 25 regarding order of operations). To check your calculator for order of operation programming, first do the following calculation by hand

and then by calculator. (Answer is at the bottom of next page. Don't look at the answer until you've finished your work.) If your calculator is not "order" programmed, don't worry—all work can still be accomplished.

$$8 + 9 \times 3 - 2 \times 14 \div 7 + 3 =$$

BASIC CALCULATOR USE FOR FIRST-TIME USERS

Example 1: $35 + 8 =$ Hint: The calculator will do exactly what you tell it to do.

1. Turn on the calculator, and enter the number 35 by pressing first 3 and then 5. The 3 first appears in the ones column, but that is only temporarily until you enter 5. The 3 then moves to its proper column and the number 35 will display.
2. Press the + button. This lets the calculator know two things: (1) The number 35 is complete and (2) It will add the next entry (or set of "order of operations" entries). The + sign will not appear.
3. Press 8. The number 35 will disappear, and the calculator is now waiting for your next instruction.
4. Press the = button. The calculator will do the math you've instructed it to do, and give you the answer to view which should be 43. It will not do that final calculation until you press the = button.

Keep your eyes on the buttons as you do your work, and check out the viewer to make sure you have entered the correct entry. If in the previous exercise you entered 36 instead of 35, the calculator will give you 44 as an answer, as it should. It will not know that you've mistakenly entered the wrong number. If you notice you entered the wrong number or instructed it to do the wrong function, tapping the CE or C button can erase your last entry (*C* stands for Clear, and *E* stands for either Entry or Error). Tapping the CE button twice will clear all entries. Sometimes the best thing to do if you've made an error is to start over completely. Some calculators may have a CA button, which stands for Clear All.

NOTES AND HELPFUL HINTS

- Little-known hazard warning (!): Calculator overuse and overreliance can lead to math tables memory loss and math skills loss!
- Math beginners: For you own benefit, you must be able to comfortably do all math calculations by hand. Don't use the calculator in the beginning.

- Experienced in math: Reacquaint yourself with calculations by hand. Check complicated procedures with calculator.
  ✓ Practicing on the calculator will lead to calculator proficiency.
- Check your own work by doing a math calculation twice, and/or even a third time.
- Some calculators may act slightly differently than described earlier.
- Important Note: It's best not to take a test using a calculator with which you're inexperienced! Some trade apprenticeship programs or job application procedures have a math entrance test, some do not. Some allow calculators, some do not. Practice on a calculator many times until you are proficient at using it correctly before bringing it to an entrance test. Occasionally I have seen students fail an entrance test because the calculator became confusing during testing. They possibly could have passed if they had done the math by hand. If you find yourself confused because of the calculator, discontinue using it and continue the test by hand. Remember, some math tests for apprenticeship or job entry may allow retesting, but only after a waiting period. This period could be a matter of days, weeks, or months, and some tests are given only once a year.

REMINDER

Every math problem in this book can be solved with the four functions: $+$, $-$, $\times$, division $\div$.
Every math problem in this book can be solved by hand.

BEST PRACTICE

Do all calculations in this book, in the open spaces, not on loose pages that can be lost. The reason for this is that you can keep this book as a reference for many years. Over time, you may forget a procedure. Seeing your own work, even though done years ago, will go a long way in reminding you of the steps you used to solve the problem.

SECTION 1

Whole Numbers

UNIT 1

Addition of Whole Numbers

Basic Principles of Addition of Whole Numbers

Whole numbers are numerical units with no fractional parts. Each whole number has a decimal point (.) at the end. The decimal point is not usually needed or used unless decimal fractions are calculated.

Addition is the process of finding the *sum* of two or more numbers. Whole numbers are added by placing them in a column with the numbers aligned on the right side of the column. Ones are lined up in their own column, tens are in their own column, and so on. Addition and subtraction calculations are begun from the ones column. The right column of numbers is added first. When whole numbers greater than 9 are added, the numbers are arranged starting with the ones lined up beneath each other. The last digit of the sum is written in the answer. The remaining digit (if any) is carried over to the next column and added. This procedure is followed until all columns are added.

EXAMPLE 1: Find this sum: 24 + 86 + 8

STEP 1: The ones column adds up to 18. The last digit of the sum, 8, is placed beneath the ones column, and the remaining digit, 1, is carried over to the tens column.

$$\begin{array}{r} \overset{1}{2}4 \\ 86 \\ +\ \ 8 \\ \hline 8 \end{array}$$

STEP 2: The tens column, including the digit 1 that was carried over, adds up to 11. Place the last digit 1 under the tens column and carry the remaining digit 1 to the hundreds column.

$$\begin{array}{r} {}^{1\,1}\\ 24 \\ 86 \\ +\ \ 8 \\ \hline 18 \end{array}$$

STEP 3: The hundreds column adds up to 1. Place 1 under the hundreds column.

$$\begin{array}{r} {}^{1\,1}\\ 24 \\ 86 \\ +\ \ 8 \\ \hline 118 \end{array}$$

ANSWER: $24 + 86 + 8 = 118$

Practical Problems

1. If the front and back sides of a foundation's walls each have a length of 48 feet and the length of each of the two end walls is 26 feet, what is the combined length of all four walls? __________

2. If a mason earns $18 per hour and a laborer earns $12 per hour, what are the total hourly labor costs for both workers? __________

3. If for a project the cost of masonry cement is $145, bricks is $445, and labor is $860, what is the total cost for all three? __________

4. A job required 600 concrete blocks for the north wall, 500 for the south wall, 800 for the east wall, and 830 for the west wall. How many concrete blocks are required for the job? __________

5. A mason lays 188 concrete blocks the first day, 212 the second day, and 200 the third day. How many blocks are laid in the three days? __________

6. Four walls of a bathroom require 32, 48, 56, and 18 square feet of glazed tile block. How many square feet of glazed tile block is required for the bathroom? __________

7. A contractor uses 38,000 bricks and orders 2,300 more to complete the job. How many bricks does the job require? __________

8. Three jobs require 420, 370, and 290 cubic yards of concrete, respectively. How many cubic yards of concrete are needed for all three jobs? __________

UNIT 2

Subtraction of Whole Numbers

Basic Principles of Subtraction of Whole Numbers

Subtraction is the process of finding the difference between two numbers. The smaller of the two numbers is placed below the larger number, keeping the right column of numbers aligned. The minus sign $(-)$ is used to indicate subtraction.

When subtracting whole numbers, it is sometimes necessary to borrow from the number in the adjacent column. When you do this, the amount borrowed must be in increments of value of the column borrowed from. Starting from the right, the first column represents units or ones, the second column represents tens, the third column represents hundreds, and so on.

EXAMPLE 1: Find the difference: $78 - 42$

STEP 1: Set up the problem aligning the columns.

$$\begin{array}{r} 78 \\ -42 \\ \hline \end{array}$$

STEP 2: Subtract the lower digit from the upper digit in the ones column. Place the answer, 6, under the ones column.

$$\begin{array}{r} 78 \\ -42 \\ \hline 6 \end{array}$$

STEP 3: Subtract the tens column. Place the answer, 3, under the tens column.

$$\begin{array}{r} 78 \\ -42 \\ \hline 36 \end{array}$$

ANSWER: $78 - 42 = 36$

EXAMPLE 2: Subtract 947 from 1244

STEP 1: Align the numbers, and begin calculating in the ones column. 7 cannot be subtracted from 4. Borrow 1 from the tens column and place in front of 4 in the ones column. Because 1 came from the tens column, you now have 14 instead of 4 in the ones column (and the tens column is reduced from 4 to 3).

Subtract the ones column.

$$\begin{array}{rrrrr} & & & ^{3} & ^{1} \\ & 1 & 2 & \not{4} & 4 \\ - & & 9 & 4 & 7 \\ \hline & & & & 7 \end{array}$$

STEP 2: Now calculate the tens column. 4 cannot be subtracted from 3, so you must again borrow, this time from the hundreds column. You now have 13 in the tens column, and 1 in the hundreds column.

Subtract the tens column.

$$\begin{array}{rrrrr} & & ^{1} & ^{13} & ^{1} \\ & 1 & \not{2} & \not{4} & 4 \\ - & & 9 & 4 & 7 \\ \hline & & & 9 & 7 \end{array}$$

STEP 3: Now calculate the hundreds column. Again, you need to borrow from the thousands column, which leaves 0 in the thousands column, and 11 in the hundreds column. Subtract the hundreds column.

$$\begin{array}{rrrr} & {\scriptstyle 11} & {\scriptstyle 13} & {\scriptstyle 1} \\ \not{1} & 2 & \not{4} & 4 \\ - & 9 & 4 & 7 \\ \hline & 2 & 9 & 7 \end{array}$$

ANSWER: 1244 − 947 = 297

Practical Problems

1. If the combined hourly wages for one mason and one laborer is $28 and the laborer earns $11 per hour, what is the mason's hourly wages? ________
2. A mason orders 56 cubic yards of concrete but uses only 47 yards. How much concrete is not used? ________
3. A mason works 48 hours but is paid only for 36 hours. For how many hours is the mason not paid? ________
4. A contractor purchases 48,000 bricks and uses 33,855 of them. How many bricks are left? ________
5. If the full length of a wall is 18 feet, and tiles will cover 6-foot length of the wall, what length of the wall is not tiled? ________
6. A contractor purchases 120 bags of cement and 20 bags of lime. The job requires 87 bags of cement and 16 bags of lime.

   a. How many bags of cement are unused? ________

   b. How many bags of lime are unused? ________

UNIT 3

Multiplication of Whole Numbers

Basic Principles of Multiplication of Whole Numbers

Multiplication builds on the principles of addition: It is helpful in figuring large quantities quickly when addition may be too slow.

The multiplication symbol ($\times$) is used to show multiplication, although other ways to show multiplication will be taught later in this book.

Each part of the problem has a name:

The top number, the one being multiplied, is called the *multiplicand*.

The lower number, the one doing the multiplying, is called the *multiplier*.

The answer is the *product*.

EXAMPLE 1: 12×8

STEP 1: Set up the problem.

$$\begin{array}{r} 12 \\ \underline{\times \ \ 8} \end{array}$$

STEP 2: Multiply the ones column ($2 \times 8 = 16$). The ones digit of the product, 6, is placed underneath the ones column, and the remaining digit 1 is carried over to the top of the tens column.

$$\begin{array}{r} {}^{1} \\ 12 \\ \times\ \ 8 \\ \hline 6 \end{array}$$

STEP 3: Multiply the tens column ($1 \times 8 = 8$). Add the number carried over (1). ($8 + 1 = 9$). Place the sum 9 under the tens column.

$$\begin{array}{r} {}^{1} \\ 12 \\ \times\ \ 8 \\ \hline 96 \end{array}$$

ANSWER: $12 \times 8 = 96$

To multiply larger numbers, first write the number to be multiplied; then write underneath it the number of times it is to be multiplied. In the following example, the number 148 is to be multiplied by 78. Write the numbers keeping the units column aligned.

EXAMPLE 2: Solve 148×78

STEP 1: Set up the problem.

Start at the far right and multiply the ones column ($8 \times 8 = 64$). Place 4 in the ones column and carry 6 to the tens column.

$$\begin{array}{r} {}^{6} \\ 148 \\ \times\ \ 78 \\ \hline 4 \end{array}$$

STEP 2: Multiply 8 by the tens column ($4 \times 8 = 32$), then add the carried number ($32 + 6 = 38$). Place the sum 8 under the tens column and carry 3 to the hundreds column.

$$\begin{array}{r} {}^{3\,6} \\ 148 \\ \times\ \ 78 \\ \hline 84 \end{array}$$

STEP 3: Multiply 8 by the hundreds column ($1 \times 8 = 8$), then add the carried number ($8 + 3 = 11$). Place the ones digit 1 below the hundreds column and carry the other 1 to the thousands column.

$$\begin{array}{r} {\scriptstyle 1\,3\,6} \\ 148 \\ \times \quad 78 \\ \hline 1184 \end{array}$$

STEP 4: Now it's time to multiply using the second number in the multiplier (7). Repeat steps 1–3 for this number, and start recording under that number (under the 7).

$$\begin{array}{r} {\scriptstyle 1\,3\,5} \\ 148 \\ \times \quad 78 \\ \hline 1184 \\ 1036 \\ \hline \end{array}$$

STEP 5: Now add the numbers recorded from multiplying each digit of the multiplier.

$$\begin{array}{r} {\scriptstyle 1\,3\,5} \\ 148 \\ \times \quad 78 \\ \hline 1184 \\ 1036 \\ \hline 11544 \end{array}$$

ANSWER: $148 \times 78 = 11{,}544$

Practical Problems

1. If a mason's pay is \$18 an hour, how much does the mason earn for a 40 hour workweek? ______
2. If each course of block is laid at a height of 8 inches, what should be the height of a block corner raised to a height of seven courses? ______
3. If a mason can lay 85 bricks per hour, how many bricks can the mason lay in an 8 hour workday? ______

4. In a certain room, each wall has an area of 432 square feet. How many square feet are in all four walls of the room? ________

5. A mason lays 29 concrete blocks in 2 hours. At this rate, how many blocks are laid in 4 hours? ________

6. A crew of stonemasons set 75 stones per hour. At this rate, how many stones are set in 40 hours? ________

7. An apprentice works 7 hours each day. How many hours does the apprentice work in a 5 day workweek? ________

8. If a bricklayer lays 595 bricks in an 8 hour workday, how many bricks could the bricklayer be expected to lay in five 8 hour workdays? ________

UNIT 4

Division of Whole Numbers

Basic Principles of Division of Whole Numbers

Division is used to determine the quantity of groups available in a given number.

Several symbols can be used to show division:

- ÷ indicates "divided by "
- The line in a fraction (¾) is called a fraction line and also indicates *divided by*. (Fractions will be taught in the following section.)
- The division box is used to do the actual work of dividing. The smaller number outside the box, which is the number doing the dividing, is called the *divisor*. The larger number inside the box, which is the number to be divided, is called the *dividend*. The answer is known as the *quotient*.

$$\begin{array}{r} \textit{quotient} \\ \textit{divisor}\overline{)\textit{dividend}} \end{array}$$

EXAMPLE 1: 96 ÷ 8

STEP 1: Set up the problem in the division box:

$8\overline{)96}$

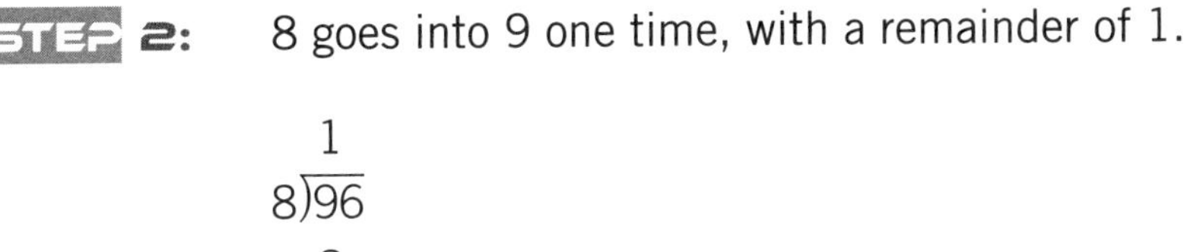

STEP 2: 8 goes into 9 one time, with a remainder of 1.

$$\begin{array}{r} 1 \\ 8\overline{)96} \\ \underline{8} \\ 1 \end{array}$$

STEP 3: Now bring down 6 and divide again. How many times does 8 go into 16?

$$\begin{array}{r} 12 \\ 8\overline{)96} \\ \underline{8} \\ 16 \\ \underline{16} \\ 0 \end{array}$$

ANSWER: 96 divided by 8 is 12.

To check your answer, multiply the divisor and the quotient:

$$\begin{array}{r} \overset{1}{1}2 \\ \underline{\times\ 8} \\ 96 \end{array}$$

Practical Problems

1. If a mason takes a total of seven bags of masonry cement to lay 875 bricks, what is the average number of bricks laid with each bag of masonry cement? __________
2. If a mason lays 240 units of 8 inch blocks in an 8 hour workday, what is the number of blocks the mason averages laying per hour? __________
3. If one cube of standard-size modular brick contains five "straps" of 102 bricks each, how many bricks are in one cube? __________

4. If there are 430 engineered-size bricks in one cube consisting of five equal "straps," how many bricks are in a single strap? ______________

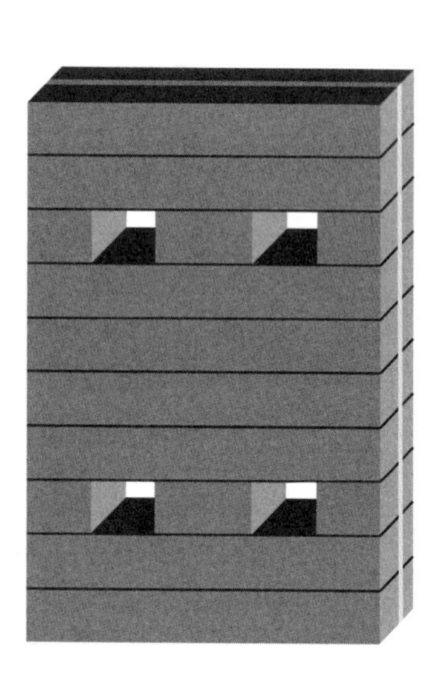

5 "STRAPS" OR "HACKS" EQUAL ONE CUBE

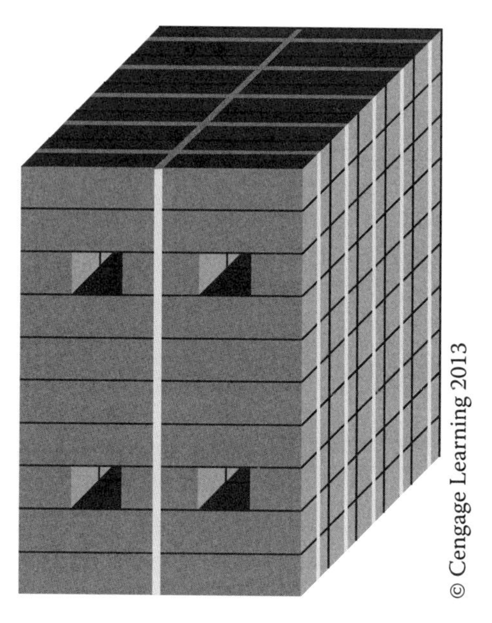

5. A mason and a helper pour and finish a 552 square-feet concrete slab in 8 hours. What is the average number of square feet of concrete they pour and finish per hour? ______________

6. Working together, four bricklayers lay 2,464 bricks in one day. What is the average number of bricks laid by each? ______________

7. If it takes seven bricks, including mortar head joints and bed joints, between them, to lay up 1 square foot of 4 inch-thick brick wall surface, how many square feet can be laid up with 763 bricks? ______________

8. If 2,934 bricks are laid in 9 hours, what is the average number of bricks laid in 1 hour? ______________

UNIT 5

Combined Operations with Whole Numbers

Basic Principles of Combined Operations with Whole Numbers

When working combined operations with addition, subtraction, multiplication, and division of whole numbers, use the following guidelines:

1. Do all operations in parentheses.
2. Solve any expressions that contain exponents or roots.
3. Multiply or divide from left to right.
4. Add or subtract from left to right.

EXAMPLE 1: $3 \times (4 - 1)$

STEP 1: Solve the parentheses.

$3 \times (3)$

STEP 2: Multiply.

$3 \times 3 = 9$

ANSWER: $3 \times (4 - 1) = 9$

EXAMPLE 2: $5 \times (6 + 2)$

STEP 1: Solve the parentheses.

$5 \times (8)$

STEP 2: Multiply.

$5 \times 8 = 40$

ANSWER: $5 \times (6 + 2) = 40$

Practical Problems

1. Two masons work a total of 560 hours on a job. Each works 8 hours per day, 5 days per week. How many weeks does each mason work? ____________
2. A mason orders 375 bags of masonry cement. On one job, he uses 186 bags, and on another, he uses 97 bags. How many bags of masonry cement are left? ____________
3. Two masons finish laying 120 square feet of concrete in 1 day. How many square feet does each mason finish laying in 5 days? ____________
4. If each bricklayer lays an average of 625 bricks per day, how many bricks can six bricklayers lay in 3 days? ____________
5. A contractor purchased 52,000 bricks. If 12,220 of these were used on one job and 36,474 were used on a second job, how many bricks did the contractor have left over? ____________
6. A contractor employs six workers. If two workers earn $22 per hour and four earn $18 per hour, what amount does the hourly wages total for these six workers? ____________

SECTION 2

Common Fractions

math-aids.com (practice)

UNIT 6

Addition of Common Fractions

Basic Principles of Addition of Common Fractions

Fractions are used to indicate quantities less than a whole. The inch is often divided into 16 equal parts. If a measurement is $\frac{5}{16}$ inch, it indicates that an inch has been divided into 16 equal parts and the measurement corresponds to 5 of these parts. Unit 25 addresses the reading of a rule or tape measurer having $\frac{1}{16}$ inch graduations.

There are two types of fractions, both of which describe less than a whole object. The object can be an inch, a foot, a mile, a ton, a bundle of welding rods, other measurements, and so forth. The two types of fractions are as follows:

1. Common fractions (fractions), for example, $\frac{1}{2}$, $\frac{3}{4}$, $\frac{5}{8}$ ________________

2. Decimal fractions (decimals), for example, 0.50, 0.75, 0.625 ________________

We will work with fractions (common fractions) in this section. Decimal fractions will be discussed in Section 3.

A fraction has two parts: the *numerator* (the number above the line, which is the pieces of a divided object) and the *denominator* (the number below the line, which is the number of pieces any one whole object is divided into). For example, if you cut a pizza into eight slices, the entire pizza is $\frac{8}{8}$, or 1. To describe three slices of this pizza, you would use the fraction $\frac{3}{8}$ (3 of 8).

Before fractions can be added, all of their denominators must be the same. Then, only the numerators are added.

EXAMPLE 1: Solve 1/8 + 3/8 + 3/8

STEP 1: These fractions have the same denominator, so you just have to add the numerators.

$$\frac{1}{8}+\frac{3}{8}+\frac{3}{8}=\frac{7}{8}$$

ANSWER: 1/8 + 3/8 + 3/8 = 7/8

If all the denominators are not the same, it is necessary to find a *common denominator.* The fractions in any problem can be changed to *equivalent fractions* with common denominators. To do this, you must find a number that divides all of the denominators of the fractions in your problem. This means that each denominator is a multiple of this number with no remainder. There will not be just one common denominator, so the *lowest common denominator* (LCD) is preferable.

It may not always be obvious what the LCD is; in such cases, a common denominator can be found by multiplying all of the denominators in the problem together. This will not likely produce the LCD, but it will produce a common denominator, which will enable the fractions to be added.

EXAMPLE 2: 1/12 + 1/3 + 1/6 + 1/4

STEP 1: Find a common denominator. All of the denominators can be divided into 12, so that is the LCD.

1/12 has the correct denominator.

Since 3 × 4 is 12, multiply ⅓ by 4/4 (which is 1) to get the LCD:

1/3 × 4/4 = 4/12

Likewise, 1/6 × 2/2 = 2/12

And 1/4 × 3/3 = 3/12

Remember: If the denominator of a fraction is multiplied by a number, you must also multiply the numerator of that fraction by the same number (see Unit 8 for more on multiplication of common fractions).

STEP 2: Now that all fractions have the same denominator, they can be added.

$$\frac{1}{12} + \frac{4}{12} + \frac{2}{12} + \frac{3}{12} = \frac{10}{12}$$

The final step in completing this problem is to reduce the fraction, if possible, to its lowest terms. If both the numerator and denominator can be divided by the same number, the fraction can be reduced. In this example, 10 and 12 are both divisible by 2.

$10 \div 2 = 5$
$12 \div 2 = 6$

ANSWER: 1/12 + 1/3 + 1/6 + 1/4 = 10/12 or 5/6

Guidelines for Reducing Fractions

If both the numerator and denominator are even numbers, or end with even numbers, both can be divided by 2. Examples: 2/8, 24/38

If the numerator and denominator end with 5, or 5 and 0, both can be divided by 5. Examples: 15/20, 35/55

If the numerator and denominator end with 0, both can be divided by 10. Examples: 30/70, 110/330

If neither of the above guidelines works for a particular fraction, experiment by dividing with 3, then 4, then 6, and so on.

Some fractions cannot be reduced, for example, when the numerator and denominator are 1 number apart. Examples: 3/4, 7/8, and 15/16

Practical Problems

1. Solve and reduce all answers to the lowest terms.

   a. $\frac{1}{16}'' + \frac{1}{16}''$ __________

   b. $\frac{1}{16}'' + \frac{1}{8}''$ __________

   c. $\frac{1}{8}'' + \frac{1}{8}''$ __________

   d. $\frac{1}{8}'' + \frac{1}{4}''$ __________

   e. $\frac{1}{4}'' + \frac{1}{16}''$ __________

   f. $2\frac{1}{4}'' + \frac{3}{8}''$ __________

2. What is the total height of three bricks that are $2\frac{1}{4}$ inches, $3\frac{1}{2}$ inches, and $3\frac{5}{8}$ inches high? __________

3. The weekly time sheet for a mason is as follows. Monday: 8 hours, Tuesday: 6 hours and 30 minutes, Wednesday: 7 hours and 20 minutes, Thursday: 5 hours and 45 minutes, and Friday: 7 hours and 15 minutes. How long did the mason work for the week? __________

Refer to the following illustration of the face of a concrete masonry unit, typically referred to as a block, to answer questions 4 and 5.

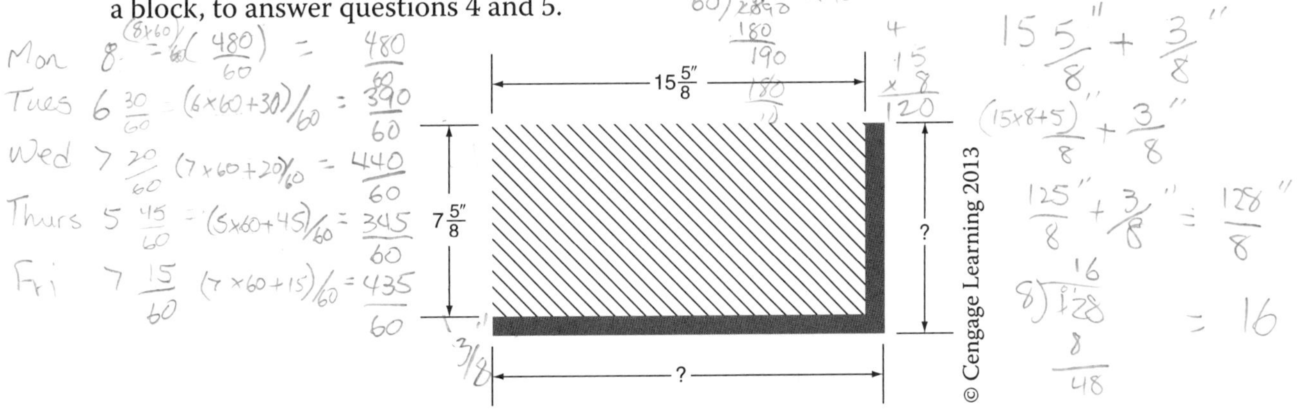

4. If mortar joints are $\frac{3}{8}$ inches wide, what is the combined length of one block and a mortar head joint? __________

5. If mortar bed joints are also $\frac{3}{8}$ inches wide, what is the combined height of one block and a mortar bed joint? __________

UNIT 7

Subtraction of Common Fractions

Basic Principles of Subtraction of Common Fractions

Subtraction of fractions works similar to addition. The first step is to make sure the fractions share a common denominator. Then subtract the numerator of the smaller fraction from the numerator of the larger fraction. Borrowing may be necessary.

EXAMPLE 1: 7/8 − 1/16

STEP 1: Find a common denominator. In this case, 16 is a common denominator.

1/16 is not changed.

$2 \times 8 = 16$, so multiply $\frac{7}{8}$ by $\frac{2}{2}$ = 14/16

STEP 2: Subtract.

$$\frac{14}{16} - \frac{1}{16} = \frac{13}{16}$$

This fraction cannot be reduced.

ANSWER: 7/8 − 1/16 = 13/16

Improper Fractions

All of the fractions discussed so far are called *proper fractions* because they properly show less than a whole object. However, a fraction is called *improper* when it represents a whole object or more. For example, $\frac{4}{4}$ is commonly expressed as 1. The fraction $\frac{7}{4}$ is also improper because it represents more than a whole (which is $\frac{4}{4}$).

You can determine whether a fraction is improper if the numerator is the same as or larger than the denominator.

If there is an improper fraction in the answer to a fraction problem, that improper fraction should be changed into its proper form, which is either a whole or a mixed number (a whole number and a fraction). Mathematically, this is done in one step by dividing the denominator into the numerator. In the preceding example, $\frac{7}{4}$ can be expressed as the mixed number $1\frac{3}{4}$.

Practical Problems

1. Solve and reduce all answers to the lowest terms.
   a. $8'' - \frac{3}{8}''$
   b. $3\frac{5}{8}'' - \frac{3}{8}''$
2. Subtract $\frac{1}{3}$ yard from $\frac{3}{4}$ yard.
3. A window frame is 5 feet high. The frame is bricked up to a height of 3 feet $1\frac{1}{16}$ inches. What is the measurement from the top of the brick to the top of the frame?
4. In an 8 hour workday, a bricklayer works two jobs. If he spends $3\frac{1}{2}$ hours on one job, how much time was spent on the second job?

5. The plan of a brick fireplace is shown. Find dimensions A, B, C, and D. ________________

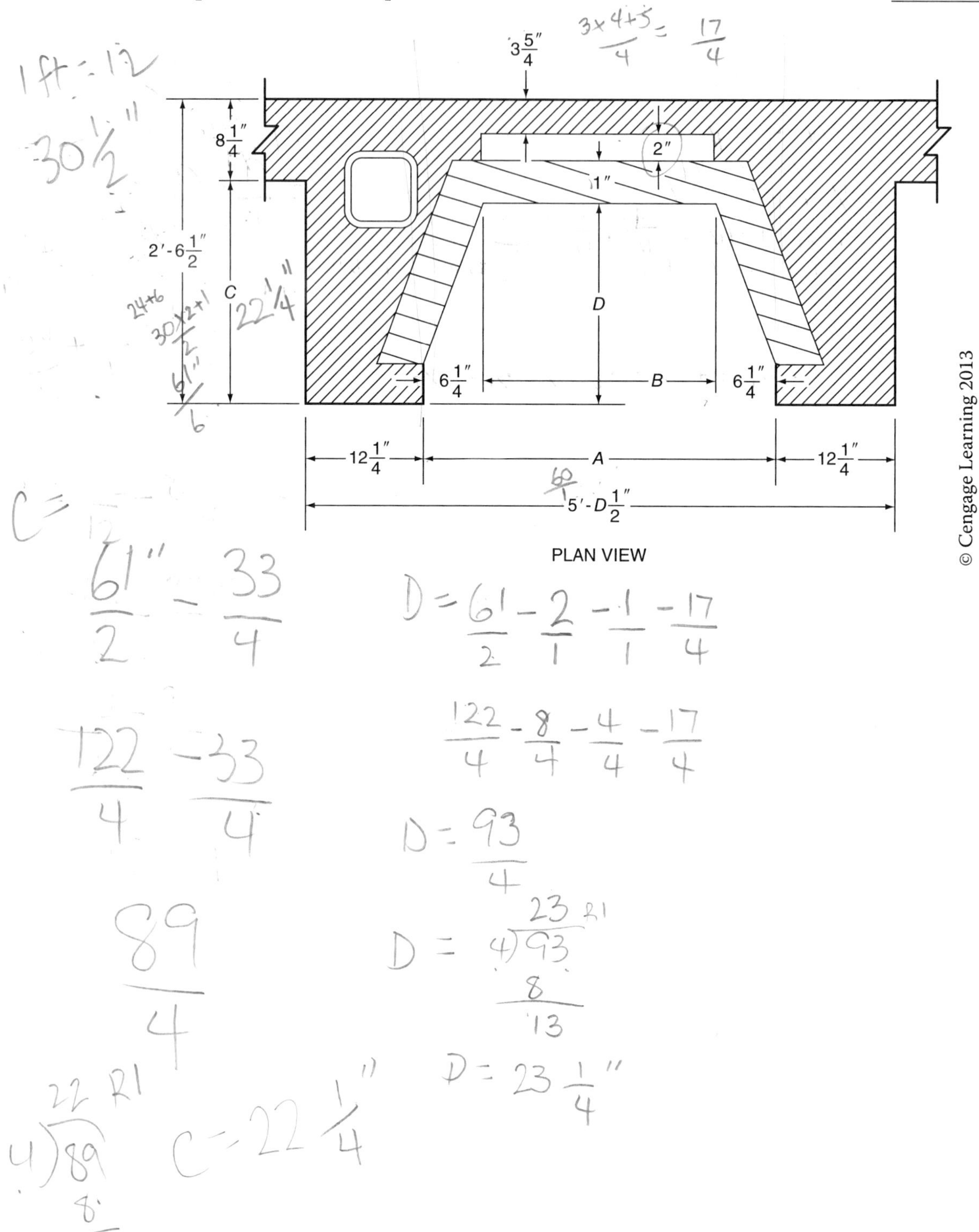

PLAN VIEW

UNIT 8

Multiplication of Common Fractions

Basic Principles of Multiplication of Common Fractions

The symbol for multiplication is "times" ($\times$). Multiplication is a short method of adding a number to itself a certain number of times. Multiplication can be shown in several ways:

$$7 \times 5 \qquad 7 \cdot 5 \qquad (7)5 \qquad 7(5) \qquad (7)(5)$$

Fractions are multiplied by multiplying the numerators together and the denominators together. To begin, all numbers must be in fraction form.

EXAMPLE 1: 5/12 × 1/7

STEP 1: Multiply the numerators together; then multiply the denominators together.

$$\frac{5}{12} \times \frac{1}{7} = \frac{1 \times 5}{12 \times 7} = \frac{5}{84}$$

When multiplying fractions, answers should always be shown in lowest terms. This fraction cannot be reduced.

ANSWER: 5/12 × 1/7 = 5/84

When fractions are multiplied, it is sometimes possible to simplify the problem by cross reduction. If a number that will divide both the numerator of one fraction and the denominator of the other can be found, the problem can be made simpler. For example, when multiplying $\frac{4}{5} \times \frac{15}{32}$, cross reduction can be used to simplify:

$$\frac{\overset{1}{\cancel{4}}}{\underset{1}{\cancel{5}}} + \frac{\overset{3}{\cancel{15}}}{\underset{8}{\cancel{32}}} = \frac{3}{8}$$

It may be necessary to change a mixed number into an improper fraction in order to multiply. To do this, multiply the whole number by the denominator and add the result to the numerator. This becomes the new numerator; the denominator is left unchanged. For example, the mixed fraction 2 ¼ can also be expressed as an improper fraction:

$$2\frac{1}{4} = \frac{(2 \times 4) + 1}{4} = \frac{8 + 1}{4} = \frac{9}{4}$$

Practical Problems

1. Solve and reduce all answers to the lowest terms.

   a. $\frac{3}{8}'' \times 6''$ ________

   b. $2\frac{5}{8}'' \times 6''$ ________

   c. $\frac{1}{3}$ yard $\times \frac{1}{5}$ ________

   d. $9\frac{5}{8}'' + 3\frac{1}{4}''$ ________

2. What is the height of 15 courses of engineered-size bricks if each course is laid to a height of 3¼ inches? ________

3. There are 17 risers in a flight of concrete steps. Each riser is 7 ¼ inches high. What is the overall height of this flight of steps? ________

4. A standard modular brick is 2¼ inches high. What is the total height of 13 standard modular bricks? ________

5. If governing regulations require extension ladders to be set at such an angle that the distance from the top support to the foot of the ladder is approximately ¼th of the working length of the ladder, how far should the base of the ladder in the following illustration be from the scaffold end frame? See the following illustration. ____________

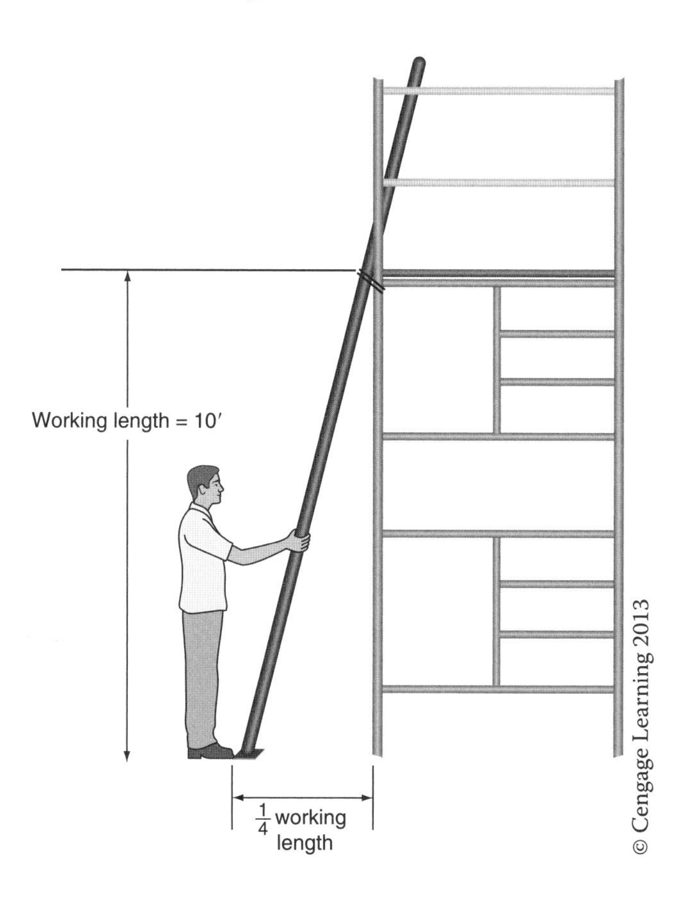

UNIT 9

Division of Common Fractions

Basic Principles of Division of Common Fractions

Division has the same steps as multiplication of fractions, with an additional step that changes division back into multiplication.

To divide common fractions, invert the divisor (the second fraction) and then multiply. (*Invert* means "turn upside down." For example, ¾ inverted is 4⁄3.) As with multiplication, all numbers must be in fraction form. To divide a whole number by a fraction, change the whole number to a fraction by giving it a denominator of 1 (e.g., 8 is equivalent to 8⁄1).

As with multiplication, it is sometimes possible to reduce fractions by cross division. The numerator of one fraction and the denominator of the other may be reduced by finding a number that divides evenly into each number.

EXAMPLE: $9/10 \div 1/3$

STEP 1: Invert the divisor and change to a multiplication problem.

$$\frac{9}{10} \times \frac{3}{1}$$

STEP 2: Multiply.

$$\frac{9}{10} \times \frac{3}{1} = \frac{9 \times 3}{10 \times 1} = \frac{27}{10}$$

Change the improper fraction to a mixed number.

$$\frac{27}{10} = 2\frac{7}{10}$$

ANSWER: 9/10 ÷ 1/3 = 2 7/10

Practical Problems

1. Solve and reduce all answers to the lowest terms.

   a. $\frac{1}{2}''$ ÷ 2 __________

   b. $\frac{1}{4}''$ ÷ 2 __________

   c. $\frac{1}{8}''$ ÷ 2 __________

   d. $\frac{3}{4}' + \frac{7}{8}'$ __________

   e. If one bag of masonry cement will lay 40 blocks, how many bags does it take to lay 100 blocks? (Give whole and fractional parts of bags.) __________

2. A double-door opening measures 8 feet 3 inches wide. What should the measurement on a ruler read at the center of this opening? __________

3. The following chart lists six modular masonry units and the number of courses of each equivalent to a height of 16 inches. Determine the height of a single course for each of these masonry units. Include fractions of inches where applicable. __________

MASONRY UNITS AND NUMBER OF COURSES IN 16 INCHES	
Concrete masonry units	2
Facing tile	3
Economy-size brick	4
Engineered-size brick	5
Standard-size brick	6
Roman brick	8

a. The height of a single course of concrete masonry units or block is _______ inches.
b. The height of a single course of facing tile is _______ inches.
c. The height of a single course of economy-size brick is _______ inches.
d. The height of a single course of engineered-size brick is _______ inches.
e. The height of a single course of standard-size brick is _______ inches.
f. The height of a single course of Roman brick is _______ inches.

UNIT 10

Combined Operations with Common Fractions

Basic Principles of Combined Operations with Common Fractions

Combined operations include addition, subtraction, multiplication, and/or division. Apply these to solve the combined operation problems. As with combined operations using whole numbers, the standard order of operation should be followed when working with combined operations of common fractions. This order is:

1. Do all operations inside parentheses. ________________
2. Solve any expressions that contain exponents or roots. ________________
3. Multiply or divide from left to right. ________________
4. Add or subtract from left to right. ________________

EXAMPLE: $(1/3 + 4/5) \times 4$

STEP 1: First perform the operation within the parentheses.

$$\frac{1}{3} + \frac{4}{5} = \frac{5}{15} + \frac{12}{15} = \frac{17}{15}$$

STEP 2: Multiply $\frac{17}{15} \times 4$.

$$\frac{17}{15} \times 4 = \frac{17}{15} \times \frac{4}{1} = \frac{4 \times 17}{15 \times 1} = \frac{68}{15}$$

ANSWER: $\left(\frac{1}{3} + \frac{4}{5}\right) \times 4 = 68/15$ or $4(8/15)$

Practical Problems

Solve and reduce all answers to the lowest terms.

1. Bricks are typically bedded in mortar joints having widths of $3/8$ inches. If contract documents specify that the acceptable tolerance for bed joints is $1/8$ more or less than $3/8$ inches, what is the maximum and minimum allowable width for bed joints?

   a. maximum: ____________

   b. minimum: ____________

Considering the dimensions of a block face to be $15\frac{5}{8}$ inches long and $7\frac{5}{8}$ inches high and assuming mortar joint width to be $3/8$ inches for both bed joints and head joints, write the correct dimension in the corresponding window when solving problems 2–4 in the following illustration.

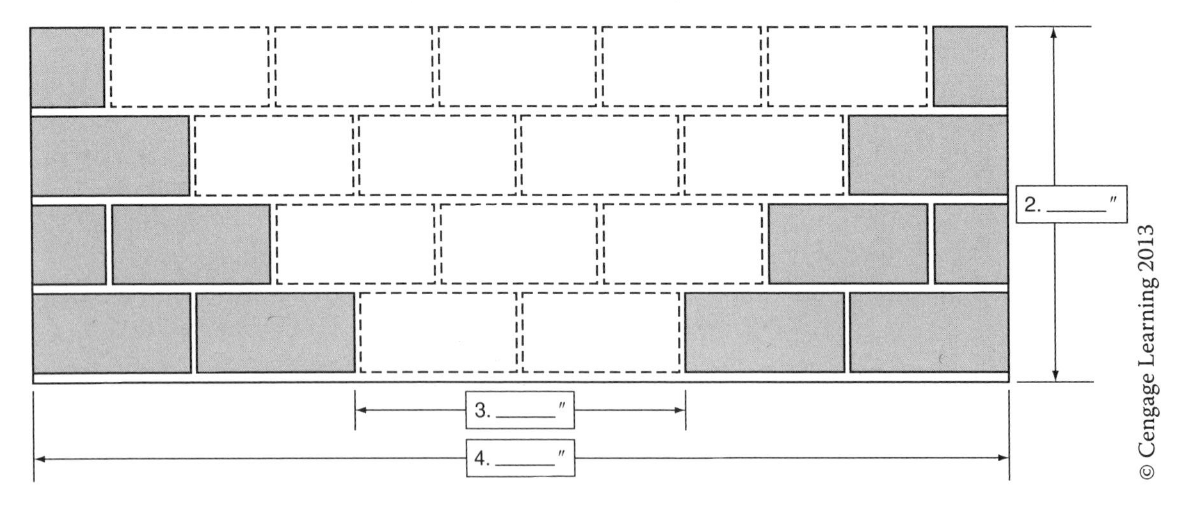

Refer to the following illustration of an 8 inch block wall, which is 83 inches long, to solve problems 5 and 6. The block at the left end of this wall alternates as being a full-length block or a half-block for every course, establishing a half-lap running bond pattern. The specified wall length prevents the block at the right end of the wall from being either a whole-length or half block. Placing these block cuts at the end of a wall ensures plumb alignment of head joints on every other course, maintaining the half-lap running bond pattern.

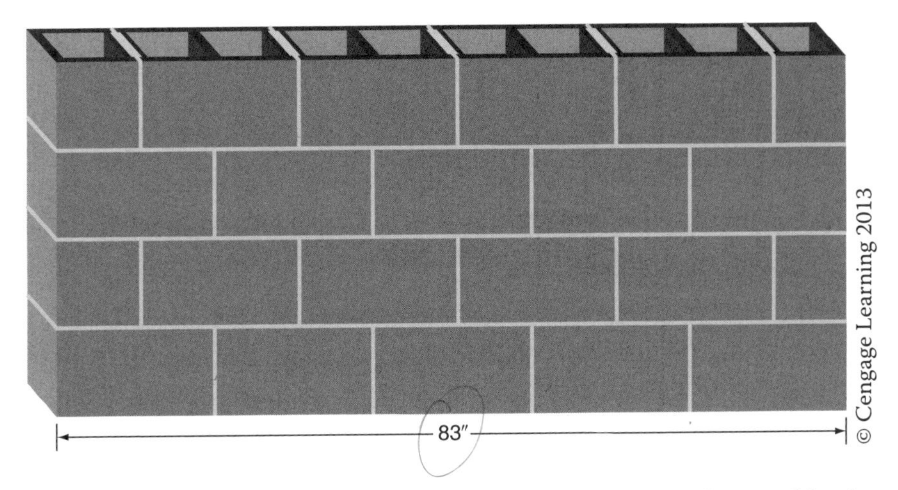

5. The following illustration shows proper placement of the cut block for courses 1 and 3. Using standard block measuring $15\frac{5}{8}$ inches long and $\frac{3}{8}$ inches wide head joints, what is the length of the cut block on courses 1 and 3 at the right end of the wall? ____________

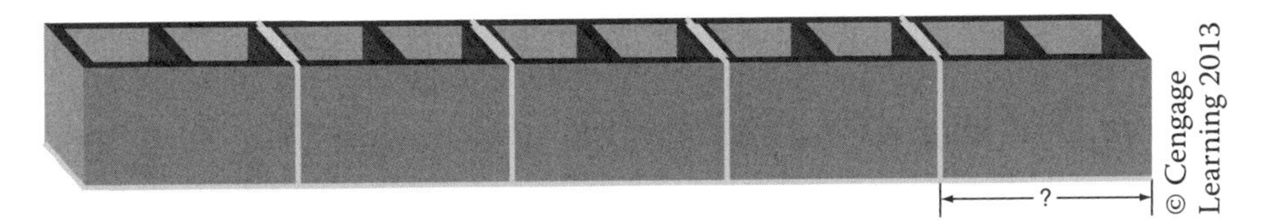

6. The following illustration shows proper placement of the block-cuts at each end of the second course. If the half-block at the left end of the wall is $7\frac{5}{8}$ inches long, each full-length block is $15\frac{5}{8}$ inches long, and each head joint is $\frac{3}{8}$ inches wide, how long is the block-cut at the right end of the wall? ______

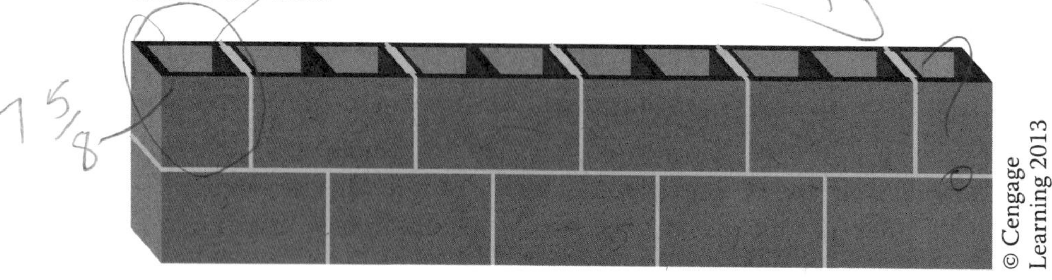

Refer to the following illustration to solve problems 7 and 8. Standard-size modular bricks with a length of $7\frac{5}{8}$ inches and a height of $2\frac{1}{4}$ inches are used to build the wall. Both head joints and bed joints are $\frac{3}{8}$ inches wide.

7. What is the length of the wall expressed as inches? ______

8. What is the height of the wall expressed as inches? ______

The following illustration represents a cross-sectional view of a wood-framed wall with an anchored brick veneer exterior wall façade and represents a typical method for exterior residential wall construction. Refer to it to solve problems 9–13.

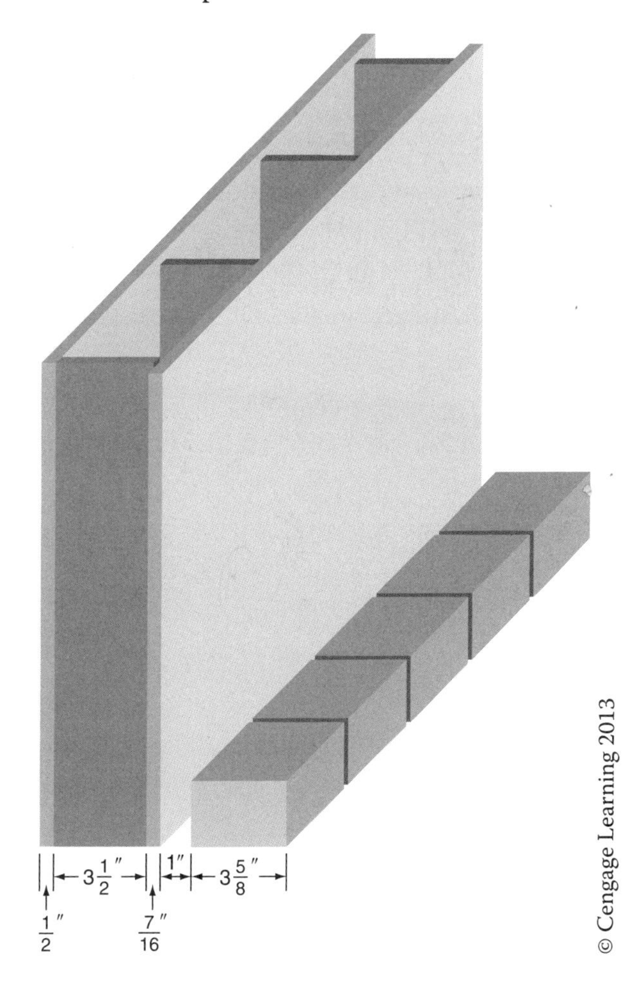

9. What is the total thickness of this illustrated wall expressed as inches? __________

10. Which has greater thickness, the interior wall board ($\frac{1}{2}''$) or the exterior wall sheathing ($\frac{7}{16}''$)? __________

11. What is the difference in thickness of these two materials? __________

12. Which has greater thickness, the 2 × 4 wall stud ($3\frac{1}{2}$″) or the brick ($3\frac{5}{8}$″)? ______

13. What is the difference in thickness of these two materials? ______

14. The following is an illustration of a hollow brick pier. It is to be constructed using standard-size bricks with length of $7\frac{5}{8}$ inches, width of $3\frac{5}{8}$ inches wide, and height of $2\frac{1}{4}$ inches. Each of the six courses is bedded in $\frac{3}{8}$ inches-wide mortar joints and all head joints are $\frac{3}{8}$ inches wide. Determine the dimensions of the pier as indicated. Solve for dimensions A–E, expressing answers to the nearest $\frac{1}{8}$ inches. ______

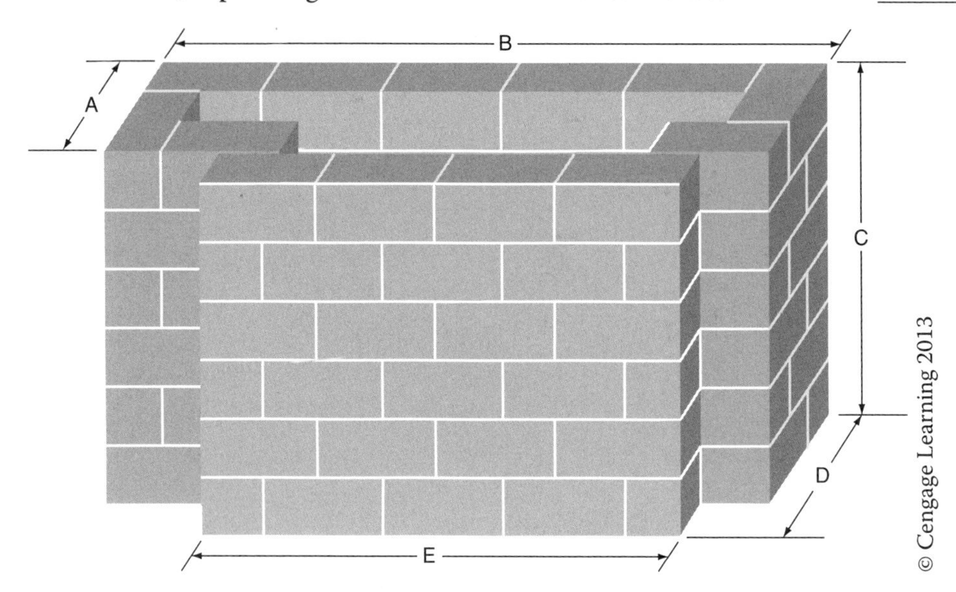

A = ________″

B = ________″

C = ________″

D = ________″

E = ________″

The following illustration shows the start of an 8 inch brick jamb. Although the length of each brick is $7\frac{5}{8}$ inches, the same as standard-size modular brick, assume the width of each brick to be less than a typical standard-size brick, at only $3\frac{3}{8}$ inches. Use the illustration to answer questions 15–17.

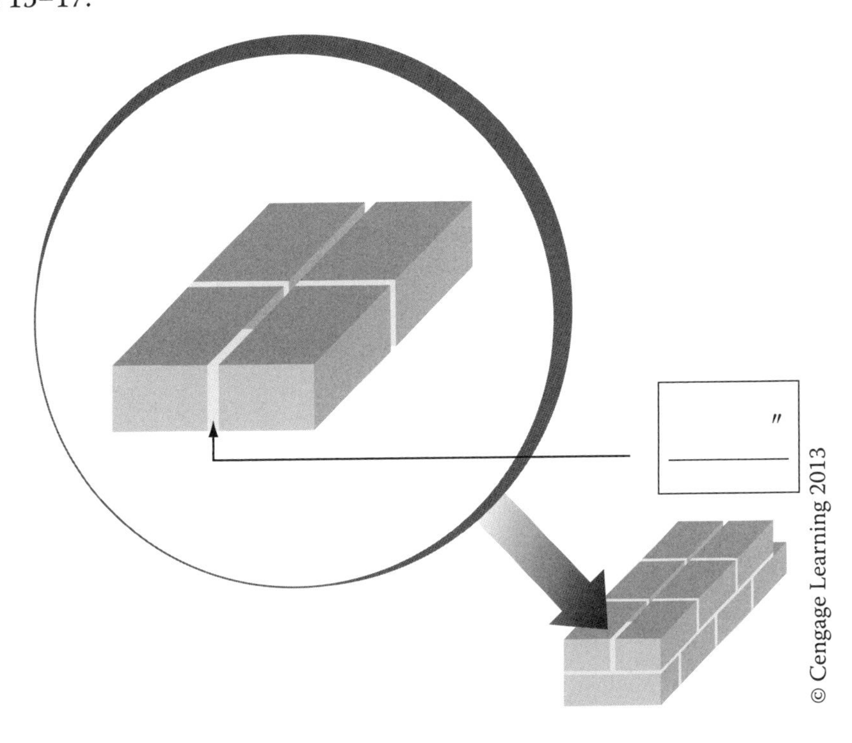

15. How much narrower are these bricks than a standard-size modular brick having a width of $3\frac{5}{8}$ inches? __________

16. What width must the head joint at the end of the wall be if both sides of the second course are to align plumb with the first course? __________

17. How much wider is this head joint than a maximum recommended joint width of $\frac{1}{2}$ inch? __________

The following illustration depicts a brick pier. The bricks are standard-size modular bricks measuring 7⅝ inches long, 3½ inches wide, and 2¼ inches high. Consider all mortar bed joints and head joints to be 5⁄16 inches wide. Solve problems 18–20.

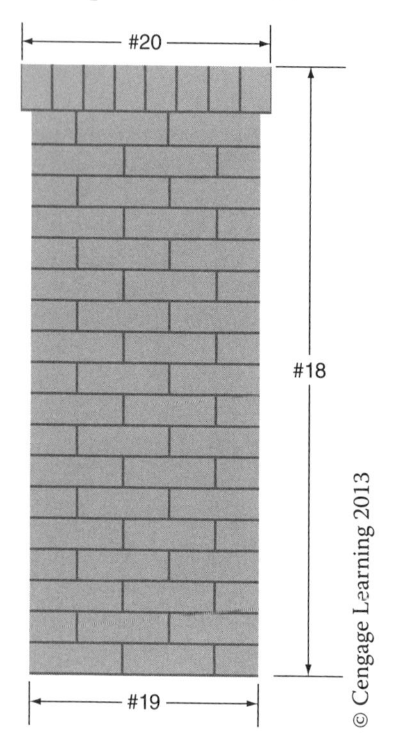

18. The combined height of all 19 courses, 18 courses laid in the stretcher position and a single rowlock course at the top, is __________

19. The length of the front side is __________

20. The length of the rowlock cap is __________

21. In masonry, the terms *pier* and *column* are sometimes used to identify the same object. But to be considered as a *column* rather than a *pier*, the definition states that the object's height must exceed four times its least lateral dimension (the lesser dimension of its length or width). Assuming the pier, in this illustration, has equal lengths on all four sides, what is the minimum number of courses of brick that would have to be laid in the stretcher position for the combined height of the stretcher courses and the rowlock cap to make it technically correct to be referred to as a *column*? __________

Refer to the following illustration of "racking a brick wall" to solve problems 22–24. Give answers to the nearest $\frac{1}{8}$ inches. Beginning on the third course, each end of the wall is racked back $\frac{5}{8}$ inches from the course below it. Assume that standard-size modular bricks ($7\frac{5}{8}$ inches long $\times$ $3\frac{1}{2}$ inches wide $\times$ $2\frac{1}{4}$ inches high) are used to build the wall and that all mortar joints are $\frac{3}{8}$ inches wide.

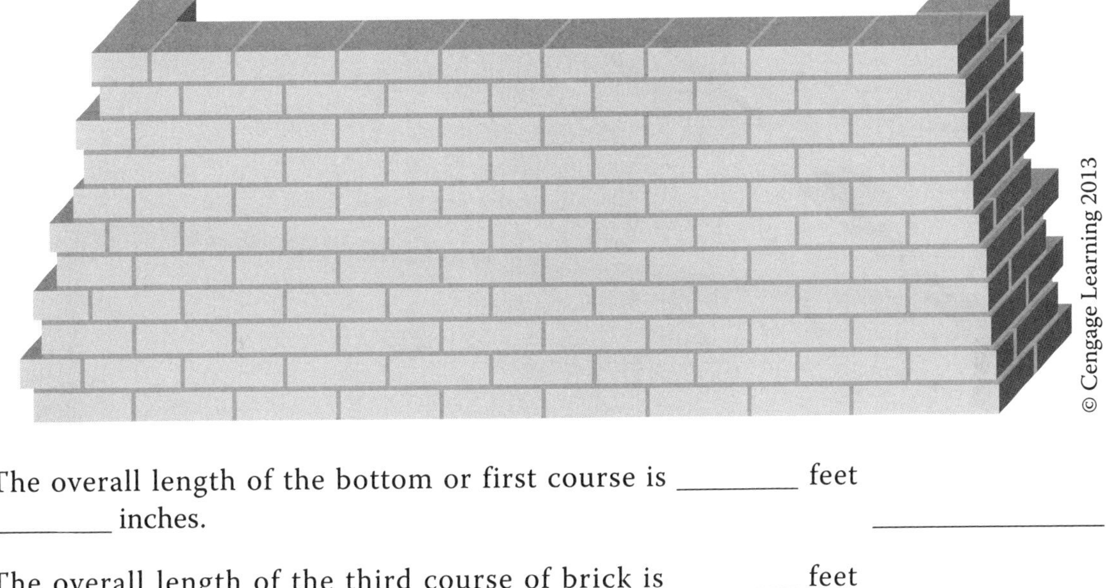

22. The overall length of the bottom or first course is ________ feet ________ inches. ________________

23. The overall length of the third course of brick is ________ feet ________ inches. ________________

24. Each end of the top course of the wall is racked back a total of how many inches less than each end of the first course? ________________

25. In the following illustration, each of the two scaffold platforms measures 7 feet long and 18 inches wide. If each is rated as having a maximum load capacity of 75 pounds per square foot, what is the maximum weight that each platform can safely support? ______

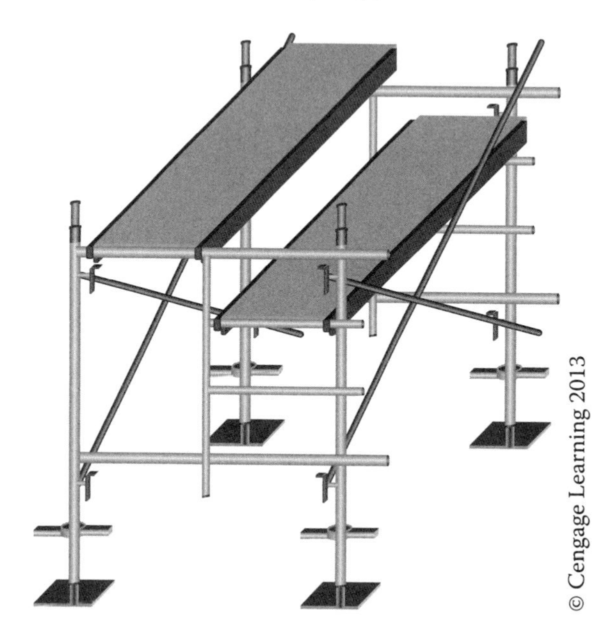

26. If governing regulations require scaffold planks to deflect, meaning to arc or bend, no more than $\frac{1}{60}$ of their supported span, what is the maximum allowable deflection for scaffold planks supported by scaffold end frames that are 7 feet apart? Give answer to the nearest $\frac{1}{8}$ inches. ______

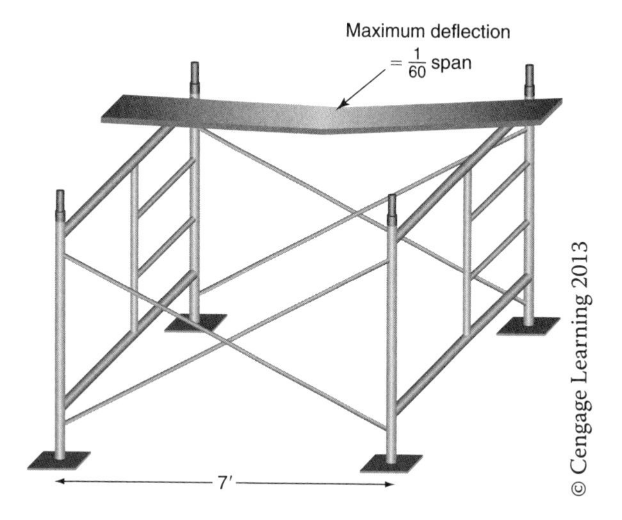

SECTION 3

Decimal Fractions

UNIT 11

Addition of Decimal Fractions

Basic Principles of Addition of Decimal Fractions

Decimal fractions are similar to common fractions in that they describe part of a whole object. In decimals, an object is divided into tenths, hundredths, thousandths, and so forth. A decimal point separates the whole numbers from the parts of a whole. Whole numbers are always to the left of the decimal point. The first place after the decimal point is called tenths, the second place is called hundredths, the third place is called thousandths, and so forth.

EXAMPLE 1: .496

Tenths	**Hundredths**	**Thousandths**
.4	9	6

Tenths describes 1 whole object divided into 10 equal parts.
Hundredths describes 1 whole object divided into 100 equal parts.

Rounding off helps express measurements according to the needs of your trade. For all decimal problems in this workbook, round off to hundredths (two places) unless otherwise noted. You may round off to three or four places if that place number is a 5 (i.e., 0.125 or 0.0625). Greater accuracy is achieved only if the final answer is rounded off, and not the numbers used to arrive at the answer.

When rounding off to tenths, if the second place number is 5 or greater, increase the tenth by 1. If the second place number is 4 or less, the tenth does not change. When rounding off to hundredths, if the third place number is 5 or greater, increase the hundredth by 1. If the third place number is 4 or lower, the hundredth does not change. When rounding off whole numbers, if the first place number is 5 or greater, increase the whole number by 1. If the first place number is 4 or lower, the whole number stays the same.

One advantage of decimal fractions over common fractions is that there is no need for a common denominator when adding or subtracting. To add decimal fractions, place the fractions in columns, aligning the decimal points.

The columns are then added like whole numbers are added. Note that each column to the left of the decimal point increases by a factor of 10 after the ones column, whereas each column to the right of the decimal point decreases by a factor of 10.

EXAMPLE 2: 4.275 + 6.5903 + 0.075 + 1.23

STEP 1: Align the decimal points in the numbers.

$$\begin{array}{r} 4.275 \\ 6.5903 \\ 0.075 \\ +\ 1.23 \end{array}$$

STEP 2: Add as with whole numbers.

$$\begin{array}{r} {}^{1\,1\,2\,1} \\ 4.275 \\ 6.5903 \\ 0.075 \\ +\ 1.23 \\ \hline 12.1703 \end{array}$$

ANSWER: 4.275 + 6.5903 + 0.075 + 1.23 = 12.1703, which can be rounded off to 12.17.

Zeros placed at the end of a decimal have no effect on its value; zeros placed in front of the decimal point have no effect on the value as long as there are no whole numbers (so .45 = 0.45 = 0.450). You can use zeros in the decimals as placeholders after the last number in the decimal. With placeholder zeros, the previous example would look like this:

$$\begin{array}{r} 4.2750 \\ 6.5903 \\ 0.0750 \\ +1.2300 \end{array}$$

Practical Problems

1. 24.50 + 16.25 + 11.75 __________
2. \$3.81 + \$4.75 __________
3. 0.75″ + 0.33″ __________
4. A bricklayer pays \$262.50 for fire bricks, \$149 for face bricks, \$44.80 for cement, and \$21.25 for sand. What is the total cost of these materials? __________
5. An apprentice works on three different jobs. The first takes 4.25 hours, the second takes 8.25 hours, and the third takes 9.75 hours. How many total hours did the apprentice work? __________
6. To finish a job, a mason pays \$92.25 for gravel, \$47.25 for sand, and \$196.80 for cement. What is the total cost of these supplies? __________
7. A contractor's estimate for a job lists \$5,750.50 for excavating and grading, \$970.25 for concrete work, \$510.75 for brickwork, and \$250.25 for setting stone. What is the total of this estimate? __________
8. A contractor pays a mason \$19.75 per hour, an apprentice mason \$12.25 per hour, and two laborers each \$10.25 per hour. What is the total hourly wages for the four workers? __________

UNIT 12

Subtraction of Decimal Fractions

Basic Principles of Subtraction of Decimal Fractions

When subtracting decimals, place the smaller number beneath the larger number, aligning the decimal points of the numbers. Then follow the procedure used for subtraction of whole numbers.

EXAMPLE 1: 24.369 − 12.93

STEP 1: Set up the problem aligning the decimal points (you may use placeholder zeros if this is helpful).

$$\begin{array}{r} 24.369 \\ -12.93 \end{array}$$

STEP 2: Subtract as with whole numbers.

$$\begin{array}{r} 2\overset{3}{\not{4}}.{}^{1}369 \\ -12.93 \\ \hline 11.439 \end{array}$$

ANSWER: 24.369 − 12.93 = 11.439, which can be rounded off to 11.44.

Practical Problems

1. 136.87 − 23.11
2. 1.45″ − 0.25″
3. 23.3 yards − 13.6 yards
4. $45.95 − $23.47
5. A mason receives $798.50 upon completion of a job. The cost of materials was $325.25. What was the mason's earnings on the job?
6. A mason has $525.75 in his company checking account. After he writes a check for $325.25, how much money is left in his account?
7. What is the height of this illustrated masonry pier?

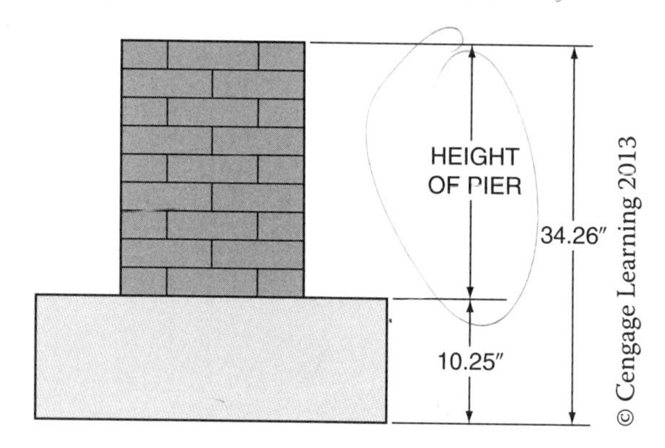

8. A bid of $24,765.50 is made on a job. If $18,068.85 is allotted for materials and labor, how much is allotted for other costs such as office and equipment expenses, insurance, and profit?

UNIT 13

Multiplication of Decimal Fractions

Basic Principles of Multiplication of Decimal Fractions

Decimals are multiplied in the same manner as whole numbers. Ignore the decimal point until the problem is solved. When you have a product, count the decimal places in each number and place the decimal point that many places from the right (the end) of the product.

EXAMPLE 1: 47.25×2.5

STEP 1: Set up the problem as with whole numbers, ignoring the decimal point.

$$\begin{array}{r} 47.25 \\ \times \quad 2.5 \\ \hline \end{array}$$

STEP 2: Multiply as with whole numbers.

$$\begin{array}{r} 47.25 \\ \times \quad 2.5 \\ \hline 23625 \\ 9450 \\ \hline 118125 \end{array}$$

STEP 3: Insert the decimal point.

The top number has two decimal places and the bottom number has one, so the decimal point will be placed three places from the right in the answer: 118.125.

ANSWER: $47.25 \times 2.5 = 118.125$, which can be rounded off to 118.13.

Practical Problems

1. 2.25″ × 2.5 __________

2. $16.74 × 12 __________

3. 7.7 × 8 __________

4. If one brick weighs 4.75 pound s, what is the weight of 128 bricks? __________

5. If one brick is 2.25 inches high, how high is a stack of 26 bricks? __________

6. Six masons each work on a job for 6.25 hours. What is the total number of hours spent on the job? __________

7. If each course of engineered-size bricks bedded in mortar is equivalent to a height of 3.25 inches, what is the total height for six courses of these bricks? (Give answer to the nearest 100th inch.) __________

8. If governing regulations require a portable emergency eye wash station on a jobsite to dispense water at the rate of 0.43 gallons of water per minute for a minimum of 15 minutes, what is the minimum amount of water that it must contain? (Give answer to the nearest 100th gallon.) __________

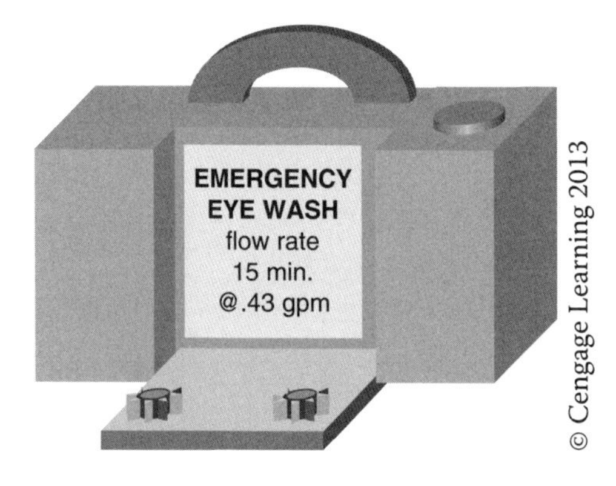

UNIT 14

Division of Decimal Fractions

Basic Principles of Division of Decimal Fractions

When decimal fractions are divided, the divisor is placed to the left of the dividend, as when dividing whole numbers. When dividing decimal fractions, however, the divisor must be a whole number, not a fraction. The divisor can be made a whole number by moving the decimal point all the way to the right of the number. The decimal point of the dividend must also be moved the same number of places to the right. The decimal point of the dividend is then placed directly above the division bracket. Now the numbers can be divided in the manner of whole numbers.

EXAMPLE 1: $4.76 \div 1.4$

STEP 1: Set up the problem in the division box.

$1.4\overline{)4.76}$

STEP 2: Move the decimal point to create a whole number for the divisor. Then place the decimal point of the dividend above the division bracket.

$$\begin{array}{r} . \\ 14\overline{)47.6} \end{array}$$

STEP 3: Now divide as with whole numbers.

```
     3.4
14)47.6
   42
    56
    56
     0
```

ANSWER: $4.76 \div 1.4 = 3.4$

Practical Problems

1. 8.4″ ÷ 4 ________________
2. 23″ ÷ 0.57 ________________
3. $98.25 ÷ 3.5 ________________
4. 5.2″ ÷ 2 ________________
5. $1,238.26 ÷ 16.50 ________________
6. When estimating a job, a contractor allows $2,600 for placing 75 cubic yards of concrete. How much is allowed for each cubic yard of concrete? ________________
7. Five masons working together earn a total of $1,268.50 on a renovation project. If each is paid an equal amount, how much is each to receive? ________________
8. A stack of bricks is 22.50 inches high. If each brick is 2.25 inches high, how many bricks are in the stack? ________________

9. If six courses of standard-size modular bricks are laid to a height of 16.5 inches, what is the height of a single course of these bricks (Give answer to the nearest 100th inch.) ____________

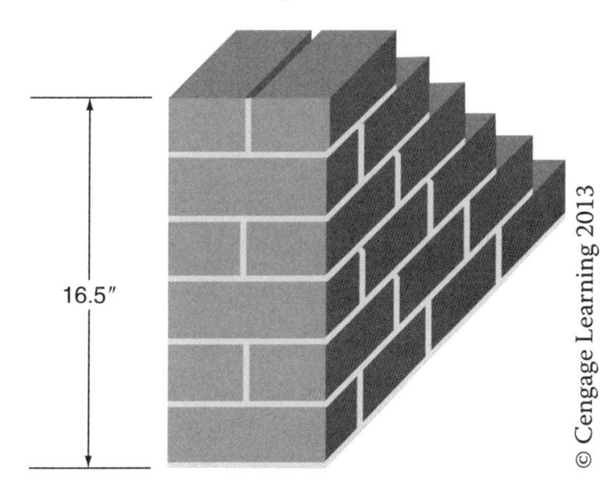

UNIT 15

Decimal Fractions and Common Fraction Equivalents

Basic Principles of Decimal Fractions and Common Fraction Equivalents

To change a common fraction to a decimal fraction, divide the numerator (the top number) by the denominator (the bottom number).

EXAMPLE 1: Change ¾ to a decimal fraction.

STEP 1: Set up the fraction in the division box.

$4\overline{)3}$

STEP 2: Divide as with whole numbers

$$\begin{array}{r} .75 \\ 4\overline{)3.00} \\ \underline{28} \\ 20 \\ \underline{20} \\ 0 \end{array}$$

ANSWER: ¾ = 0.75

To change a decimal fraction to a common fraction, note that the number of decimal places determines the number of zeros added to the number 1 to create the denominator. The decimal number becomes the numerator.

EXAMPLE 2: Change .35 to a common fraction

STEP 1: .35 has two decimal places, so the denominator will have two zeros.

$$\frac{35}{100}$$

ANSWER: $.35 = {}^{35}\!/_{100}$

Practical Problems

Express the following common fractions as decimal fractions.

1. $\frac{5}{8}$ ______
2. $\frac{1}{2}$ ______
3. $\frac{5}{16}$ ______

Express the following decimal fractions as common fractions.

4. 0.1875 ______
5. 0.3125 ______
6. 0.4375 ______
7. The actual height, in inches, of a standard-size brick is $2\frac{1}{4}$ inches. Express this height as a decimal fraction. ______
8. A caulked expansion joint between a door frame and surrounding brickwork is 0.25 inches wide. Express this as a common fraction. ______
9. The weight of a tile is 2.75 pounds per square foot. Express this weight as a common fraction. ______
10. A mortar joint measures $\frac{3}{8}$ inches. Express this thickness as a decimal fraction. ______

UNIT 16

Combined Operations with Decimal Fractions

Basic Principles of Combined Operations with Decimal Fractions

Some problems combine addition, subtraction, multiplication, and/or division of decimal fractions.

As with whole numbers and common fractions, the following order of procedures must be followed for decimal fractions as well:

1. Carry out all operations inside parentheses.
2. Solve any expressions that contain exponents or roots.
3. Multiply or divide from left to right.
4. Add or subtract from left to right.

EXAMPLE 1: $(0.25 + 3.6) \times 2.5$

STEP 1: Solve the parentheses.

$$\begin{array}{r} 0.25 \\ +3.6 \\ \hline 3.85 \end{array}$$

STEP 2: Multiply 3.85×2.5

$$\begin{array}{r} 3.85 \\ \times\ 2.5 \\ \hline 1925 \\ 770 \\ \hline 9625 \end{array}$$

Place the decimal point: The two numbers have three decimal places: 9.625

ANSWER: $(0.25 + 3.6) \times 2.5 = 9.625$

Practical Problems

1. $4.2 \times (6 - 3.5)$ ______
2. $9.858 - (4.25 + 2.5)$ ______
3. $6 + (4.25 - 1.97)$ ______
4. $2.457 \times (3.29 + 1.576)$ ______
5. If a portable eye wash station, weighing 21.5 pounds when empty, is filled with 7.5 gallons of water weighing 8.3 pounds per gallon, what is the weight of the eye wash station and water combined? ______
6. If 15,000 bricks are required to brick veneer a house, how much time can the mason save if the average time for laying each brick is reduced by 10 seconds? (Give answer to the nearest hundredth hour.) ______
7. Referring to the preceding problem (6), if the mason earns \$18.50 an hour and the laborer earns \$10.75 an hour, what is the dollar amount of savings to the contractor for the increased productivity? ______

8. If a concrete masonry lintel measuring 8 inches × 8 inches × 4.5 feet weighs 252 pounds, what is the weight per linear foot of such a lintel? ______________

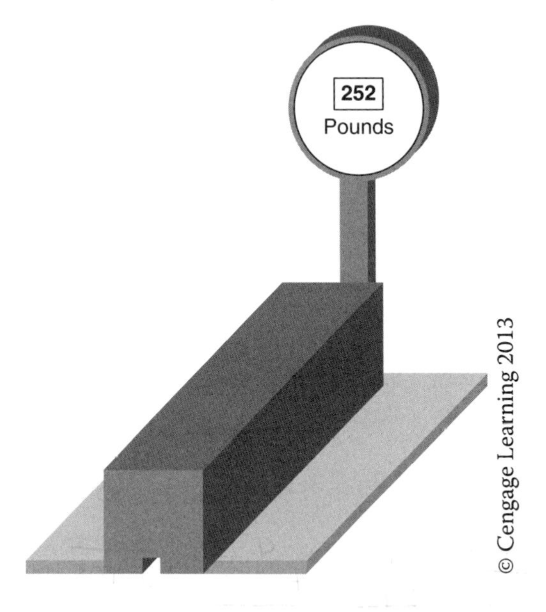

SECTION 4

Percentages, Interest, Averages, and Proportions

UNIT 17

Calculating Percentages

Basic Principles of Percentage

Percentages are used to express a part or a portion of a whole. *Percent* means hundredth or number per 100. Percentages are based on the principle that 100% represents a whole, 50% represents ½, 25% represents ¼, and so forth. They are written with the percentage symbol (%), which takes the place of two decimals.

If a mason has 100 bags of cement, and uses 52 of them on a project, the mason used 52% of the bags of cement.

To calculate percentages (%):

a. Formulate a fraction from the information given.
b. Change the fraction into a decimal.
c. Change the decimal into a percentage.

EXAMPLE 1: Calculate the percentage used when 10 out of 20 bags of cement are used on a job.

STEP 1: Formulate a fraction from the information given.

$$\frac{10}{20}$$

This fraction reduces to

$$\frac{1}{2}$$

STEP 2: Change the fraction to a decimal.

Divide the numerator by the denominator:

$$2\overline{)1.0}$$ = 0.5

$$\begin{array}{r} 0.5 \\ 2\overline{)1.0} \\ \underline{1.0} \\ 0 \end{array}$$

STEP 3: Change the decimal into a percentage.

Move the decimal point two places to the right (multiply by 100), and add the % symbol at the end of the number.

$0.5 \times 100 = 50$

$= 50\%$

Note that the decimal point is not shown if it is at the end of the number.

When 10 of 20 bags of concrete mix are used, 50 of the concrete mix is used.

Finding What Percentage One Number Is of Another

To find out what percentage one number is of another, first change the numbers to a fraction and then change the fraction to a percentage.

EXAMPLE 1: 30 is what percentage of 50?

$$\frac{30}{50} = 0.60 = 60\%$$

W = P / %

P = % x Whole

Practical Problems

1. 50% of 100 ______
2. 25% of 64 ______
3. 33% of 27 ______
4. 5% of 250 ______
5. 10% of 396 ______
6. 4.5% of 995 ______
7. It is calculated that a house requires 22,500 bricks to complete its brick veneer facade. The contractor's experience indicates that it takes 4% more than the calculated amount to complete a job because the remaining pieces of bricks for making cuts around window and door openings are not utilized. If this job requires 4% more than the calculated amount due to brick waste, how many more bricks than the available 22,500 will the contractor take to complete the job? ______
8. If brick veneer for a house facade requires either 22,500 standard-size bricks or 18,750 engineered-size bricks, what percentage of fewer bricks are needed if engineered-size bricks are used? ______
9. Two masons build a house foundation having a total of 2,400 blocks. The more experienced mason lays 1,440 blocks and the other mason lays 960. What percentage of the total did the more experienced mason lay? ______
10. The local building supply store bills a masonry contractor the amount of $3,750.46 for the previous month's purchases. If the masonry contractor makes payment within 30 days of the billing period, she can deduct 5% from the total billing. How much money can she save by making payment within the 30 days? ______
11. A masonry contractor notices that many of the 1,800 blocks being delivered to his jobsite have small but noticeable surface chips. After examining each block, he finds 216 such blocks. The supplier informs him that production specifications permit as many as 5% of the blocks to have these appearance imperfections. What percentage of the blocks has noticeable chipped surfaces? ______

12. Reports suggest an increase of as much as 23% in job productivity when corner poles or masonry guides are used rather than having masons build brick leads at the ends of walls. If a crew of four masons averages laying 12,000 bricks each 40 hour workweek and rely upon building brick leads, how many bricks might they lay in the same time period if corner poles are used, eliminating the building of brick leads? __________

13. The maximum spacing for corrugated wall ties anchoring brick veneer to structural back-up walls is one wall tie for every 2.67 square feet. A common practice is to fasten a wall tie to each wall stud spaced 16 inches apart along the entire length of walls. Ties are typically placed every five courses of engineered-size bricks or six courses of standard-size bricks, a height interval approximately 16 inches, a spacing of one wall tie every 1.77 square feet. What percentage does this spacing exceed the minimum requirement of one wall tie per 2.67 square feet? __________

14. What percentage of a square foot is a 16 inches × 8 inches foundation vent? __________

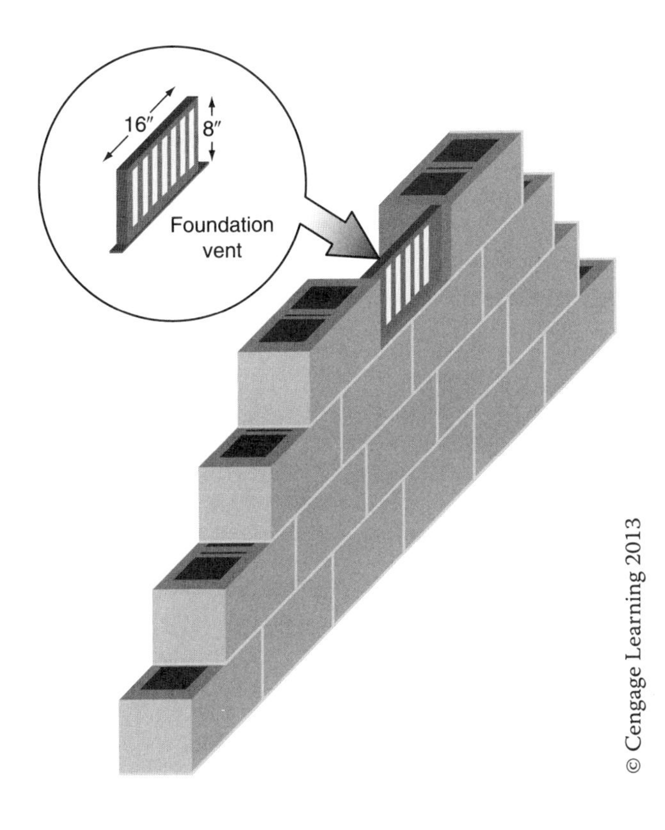

UNIT 18

Interest

Basic Principles of Interest

Interest is a way of using percentage. When money is borrowed, the amount borrowed is known as the *principal*. The amount charged for the use of the borrowed money is the *interest*, and the rate of interest is expressed as a percentage. Interest is usually compounded on a yearly basis (per annum). When the term of the loan expires, the money repaid is the principal and the interest.

EXAMPLE 1: A company borrows $18,000 for a period of 1 year at a rate of 12%. How much money is repaid at the end of the year?

STEP 1: Change the interest rate into a decimal fraction.

$12\% = 0.12$

STEP 2: Multiply the principal by the decimal fraction to determine the amount of interest paid.

$\$18{,}000 \times 0.12 = \$2{,}160$

STEP 3: Add the interest paid to the principal to determine the total amount paid.

$\$18{,}000 + \$2{,}160 = \$20{,}160$

ANSWER: The total amount repaid is $20,160.

If the duration of the loan is longer or shorter than 1 year, divide the annual interest by 12 months and multiply by the number of months.

EXAMPLE 2: Using the preceding example, determine the amount of money repaid on this loan if the loan period was extended to 18 months.

STEP 1: Determine the annual interest.

$\$18{,}000 \times 0.12 = \$2{,}160$

STEP 2: Divide the annual interest by 12 to determine the monthly interest.

$\$2{,}160 \div 12 = \$180$

STEP 3: Multiply the monthly interest by the number of months to determine total interest paid.

$\$180 \times 18 = \$3{,}240$

STEP 4: Add the interest to the principal for the total amount repaid.

$\$18{,}000 + \$3{,}240 = \$21{,}240$

ANSWER: The total amount repaid is $21,240.

Practical Problems

1. A contractor borrows $6,500 for 1 year at an interest rate of 18%. What is the total amount repaid at the end of the year? ________
2. A contractor borrows $12,000 at an interest rate of 7% per annum. If the loan is repaid in 8 months, what is the total amount repaid? ________
3. A mason borrows $3,600 for supplies at an interest rate of 12% per annum. If the loan is repaid in 1 year, what is the total interest charged? ________
4. A contractor borrows $23,000 at an interest rate of 14% per annum. What is the monthly interest on this loan? ________
5. A mason borrows $2,500 at an interest rate of 9% per annum. If the loan is repaid 16 months later, what is the total amount repaid? ________

UNIT 19

Averages

Basic Principles of Averages

Averaging figures can give the mason helpful information about a variety of subjects. Average value is found by adding individual values of some number of units and dividing the sum by the total number of units.

EXAMPLE 1: Find the average weekly pay for Jill, a mason at Warner Stonemasons during the month of March.

Week 1, Jill made \$684.00
Week 2, \$731.00
Week 3, \$825.00
Week 4, \$736.00

STEP 1: Add the figures together.

$$\begin{array}{r} \scriptstyle 1\,1 \\ \$684.00 \\ 731.00 \\ 825.00 \\ +\ 736.00 \\ \hline \$2976.00 \end{array}$$

STEP 2: Divide the sum by the number of figures.

```
    744
 4)2976
   28
   ---
    17
    166
    ---
     16
     16
     --
      0
```

ANSWER: Jill averaged $744.00 per week in March.

Practical Problems

1. A contractor used 12,500 bricks for one job, 14,375 for a second job, and 9,770 for a third job. What is the average number of bricks used per job? __________

2. A mason worked on a project for 6 weeks. Her weekly hours were as follows: 37 hours, 42 hours, 32 hours, 28 hours, 34 hours, and 39 hours. What was this mason's average weekly hours? __________

3. In the course of his work, a mason drove the following mileage:

   Monday: 56 miles
   Tuesday: 78 miles
   Wednesday: 39 miles
   Thursday: 64 miles
   Friday: 68 miles

   What was the average number of miles driven per day? __________

4. A contractor pays three employees each $18 per hour and two employees each $24 per hour. What is the average hourly rate of pay for these employees? ____________

5. A masonry contractor had the following supply bills in June:

   Week 1: $1,375

   Week 2: $4,140

   Week 3: $979

   Week 4: $1,267

   What was the contractor's weekly average expenditure for supplies? ____________

6. One mason lays 160 concrete blocks the first day, 198 concrete blocks the second day, and 215 concrete blocks the third day. How many concrete blocks did the mason lay each day, on average? ____________

UNIT 20

Calculating Ratios and Proportions

Basic Principles of Ratios

Ratios are a way of comparing two numbers. In most cases, a ratio is a comparison of two whole numbers. Ratios, like fractions, are generally reduced to their lowest terms. Ratios are often expressed like this: 16:8 (read as 16 to 8). Ratios can also be written in fractional form. The previous ratio can be written as $\frac{16}{8}$. An *inverse ratio* is written in reverse order of the original ratio. The inverse ratio of 16:8 is 8:16.

EXAMPLE 1: Express 25:40 (also expressed $\frac{25}{40}$) in its lowest terms.

STEP 1: Divide each number in the ratio by the largest common factor.

$$\frac{25 \div 5}{40 \div 5} = \frac{5}{8}$$

ANSWER: 25:40 expressed in lowest terms is 5:8 ($\frac{5}{8}$).

EXAMPLE 2: Express $\frac{3}{8} : \frac{5}{8}$ in its lowest terms.

STEP 1:

$$\text{Rewrite } \frac{\frac{3}{8}}{\frac{5}{8}} = \frac{3}{8} \times \frac{8}{5} = \frac{3}{5} = 3:5$$

ANSWER: $\frac{3}{8}:\frac{5}{8}$ in its lowest terms is 3:5.

Just so You'll Know!: A Lesson in Math and Science

1 + 3 does not always equal 4. Mixing 1 cubic foot of masonry cement with 3 cubic feet of sand does not necessarily produce 4 cubic feet of mortar. Typically it is much less. How can this be so? Masonry sand typically contains 10% or more water by volume. The water molecules on the sand particles push the grains of sand apart and increase the volume of the sand as much as 30%, a condition known as *sand bulking.* As water is added to the ingredients during mixing, sand bulking is eliminated.

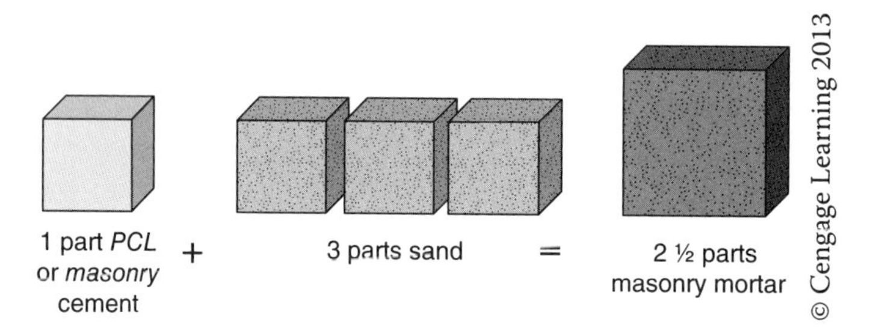

For accurate proportioning of sand with masonry cement, dampen dry sand before mixing. Mortar has too much sand content if it is made with sand too dry. When sand is too wet, the resulting mortar has too little sand content. To produce mortar having the intended strength and workability, masonry sand should form a clump in the palm of the hand when squeezed with the fingers and released.

Basic Principles of Proportions

Proportion is the equality of two ratios. For example, 3:4 = 9:12.

There are two basic methods of solving problems dealing with proportion. One method is to multiply the *means* together and the *extremes* together. The means are the two inside digits, and the extremes are the two outside digits.

EXAMPLE 1: 3:15 = 4:?

STEP 1: Multiply the means together and the extremes together.

$3x = 60$

STEP 2: To find the value of x, divide both sides by 3.

$3x/3 = x$

$60/3 = 20$

$x = 20$

ANSWER: 3:15 = 4:20

Another way to solve problems of proportion is to cross-multiply. When cross-multiplication is used, the two ratios are written as fractions separated by an equal sign.

EXAMPLE: 2:3:15 = 4:?

STEP 1: Rewrite the ratio as fractions.

$$\frac{3}{15} = \frac{4}{x}$$

STEP 2: The top digit of one ratio is then multiplied by the bottom digit of the other ratio.

$$\frac{3}{15} \times \frac{4}{x}$$

$3x = 60$
$x = 20$

Solve for x.

ANSWER: 3:15 = 4:20

Practical Problems

1. 5:1 is the same as 10:________
2. 6:12 × 9:________
3. 1:2¼ as ________:18
4. $\frac{4.5}{1} = \frac{?}{6}$
5. $\frac{?}{125} = \frac{8}{1000}$
6. $\frac{40}{1} = \frac{2500}{?}$
7. $\frac{22000}{?} = \frac{125}{1}$
8. If masonry cement and sand are combined in a ratio of 1:3, respectively, to make mortar, how many cubic feet of sand are mixed with a single 1-cubic foot bag of masonry cement?
9. If the ratio of standard-size brick to engineered-size brick is 5:6 for building a wall, how many engineered-size bricks would it take to build the same wall requiring 18,000 standard-size bricks?
10. If it takes 4½ paving bricks to complete 1 square foot of brick pavement, how many bricks are needed to complete a 360 square-foot brick patio?
11. The following illustration shows the roof pitch on a house to be 5⁄12, meaning that for every 12 inches of horizontal length, the roof surface is pitched or raised 5 inches. If the length of the gable end of the house is 28 feet, what is the total rise of the gable?

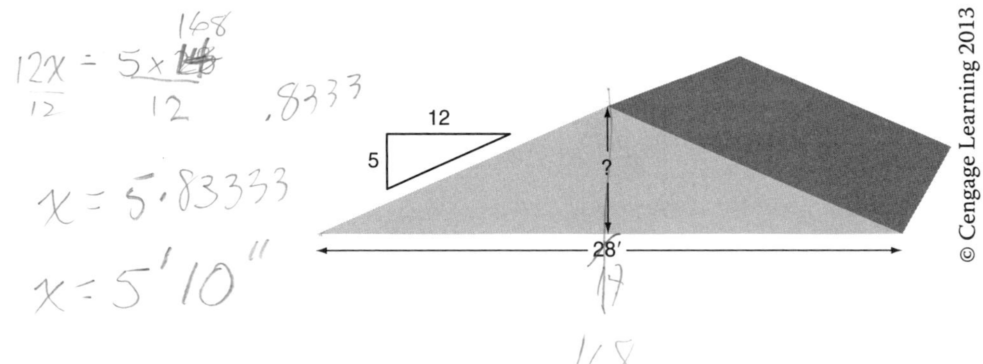

12. If governing regulations require brick veneer walls to be anchored to structural framing by using a minimum of one wall tie per 2.67 square feet, what is the minimum number of wall ties needed for a house requiring 2,100 square feet of brick veneer? __________

13. A mason calculates that it will take 5,000 bricks to complete a patio. A similar job completed by the mason required 3,500 bricks and took him 56 hours to complete. Based on the previous job, how many hours should the mason take to complete this job? 80 hours

14. A typical gasoline engine for a mortar mixer is equipped with a 6:1 speed reduction gear. If the gasoline engine without the speed reduction gear is rated at 3,200 rpm (revolutions per minute), at what rpm does the engine shaft turn with the 6:1 speed reduction gear attached? __________

15. If governing regulations require bracing-supported scaffolding at no more than a 4:1 ratio (height to scaffold-end frame width) to prevent tip-over, at what height must supported scaffolding be braced to prevent tip-over when using 5 foot wide scaffold end frames? 20

16. A 4:1 safety factor is required for anchors securing brick veneer to back-up walls. If it is calculated that installed wall anchors can experience a force of 95 pounds of compression due to the forces of wind against the face of a wall, what should be the minimum rating for the wall anchors? __________

17. How many square feet of outside air ventilation must be supplied to a foundation crawl space having an area of 1,350 square feet if a minimum of 1 square foot of ventilation is required for each 150 square feet of area? 9 ft²

13. $56 : 3500 = x : 5000$

$\frac{3500x}{3500} = \frac{56 \times 5000}{3500}$

$x = 80$

15. $4 : 1 = x : 5$

$\frac{1x}{1} = \frac{4 \times 5}{1}$

$x = 20$

$1 : 150 = x : 1350$

$\frac{150x}{150} = \frac{1 \times 1350}{150}$

$x = 9$

18. If Portland cement, sand, and gravel are mixed in a ratio of 1:2:3, respectively, to make concrete, how many cubic feet of sand are added to 27 cubic feet of Portland cement? How many cubic feet of gravel are added to this mixture? ______________

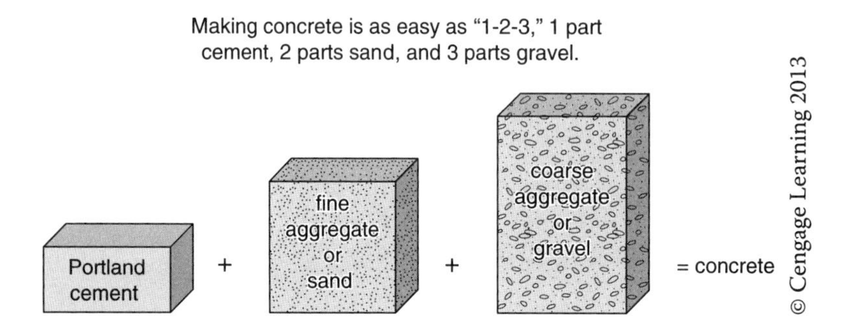

19. A particular job specifies that ⅝-inch-diameter steel reinforcement rods, also known as #5 rebars, are to extend from the bottom to the top of 8-foot tall CMU foundation walls, each placed in grout-filled block cells. The first rebar is placed at one end of the wall and all others are spaced 4 feet apart along the length of the wall. How many 8-foot long rebars are needed for a wall 72 feet long? ______________

SECTION 5

Powers and Roots

UNIT 21

Powers

Basic Principles of Powers

The *power* of a number is also known as its *exponent.* If a number is squared, for example, it is raised to the second power, or has an exponent of 2. The exponent is the small number written above, to the right of the number, as shown at the end of this paragraph. This means that the number is to be multiplied by itself that many number of times. Powers, then, are used to simplify the representation of the repeated multiplication of a number with itself. If a number is cubed, it means that the number is raised to the third power, or has an exponent of 3. In the expression 4^3, 3 is the power or exponent of 4.

4^3 simply means to multiply 4 with itself three times (this is not the same as 4×3). Without using powers, 4^3 must be written as (4)(4)(4) or $4 \times 4 \times 4$. The product is 64; so $4^3 = 64$.

A number can be raised to any power by using it as a factor that number of times.

EXAMPLE 1: Raise 5 to the fourth power.

STEP 1: $5^4 = 5 \times 5 \times 5 \times 5$

STEP 2: Multiply.

$(5 \times 5)(5 \times 5) = 25 \times 25 = 625$

ANSWER: $5^4 = 625$

Practical Problems

Raise the following expressions to the powers indicated.

1. $2^2 =$ ________
2. $3.5^2 =$ ________
3. $4^2 =$ ________
4. $10^2 =$ ________
5. $7^2 =$ ________
6. $2^3 =$ ________
7. $5^3 =$ ________
8. $3.75^3 =$ ________
9. $3^3 =$ ________

UNIT 22

Roots

Basic Principles of Roots

The root of a number is the opposite or inverse of its power. The square root of a number is determined by finding which number used as a factor twice will equal the given number. The square root of 9 is the number that, when squared, is 9. The square root of 9 is 3 because 3 squared (3×3) equals 9.

When the root of a number is to be found, the number is placed under a *radical sign* ($\sqrt{\ }$). If some root other than a square root is to be found, a small number is placed outside the radical sign. This number indicates what root is to be found.

The two roots frequently found in mathematical formulas and equations used in masonry are *square roots* and *cube roots*. The symbol representing cube root is $\sqrt[3]{\ }$. Since $2 \times 2 \times 2 = 8$, the cube root of 8 is 2.

EXAMPLE 1: $\sqrt{36}$

STEP 1: $6 \times 6 = 36$

ANSWER: $\sqrt{36} = 6$

EXAMPLE 2: $\sqrt[3]{125}$

STEP 1: $(5 \times 5) \times 5 = 25 \times 5 = 125$

ANSWER: $\sqrt[3]{125} = 5$

Practical Problems

Use a calculator to find the root of each of these numbers. Round off the answers to the nearest thousandth (three places to the right of the decimal).

1. $\sqrt{2}$ = 1.414
2. $\sqrt{64}$ = 8×8 = 64 8
3. $\sqrt{25}$ = 5×5 = 25 5
4. $\sqrt{36}$ = 6×6 = 36 6
5. $\sqrt{144}$ = 12×12 = 144 12
6. $\sqrt{225}$ = 15×15 = 225 15
7. $\sqrt{23}$ = 4.796
8. $\sqrt[3]{64}$ = ____________
9. $\sqrt[3]{27}$ = ____________
10. $\sqrt[3]{1{,}000}$ = ____________

UNIT 23

Combined Operations with Powers and Roots

Basic Principles of Combined Operations with Powers and Roots

This unit provides practical problems involving combined operations with powers and roots. Powers, or exponents, are a convenient way to indicate the number of times a quantity is to be multiplied by itself. The root of a number is the inverse of its power.

The Pythagorean Theorem

The Pythagorean Theorem is a mathematical formula containing powers that can be used to lay out and confirm the square alignment of a structure's two adjacent walls. Two adjacent walls are said to be aligned *square* when they form a 90° angle. The theorem can be written as an equation relating the lengths of the sides a, b, and c.

Where a and b represent the two adjacent sides creating a right angle and c represents the hypotenuse:

$$a^2 + b^2 = c^2$$

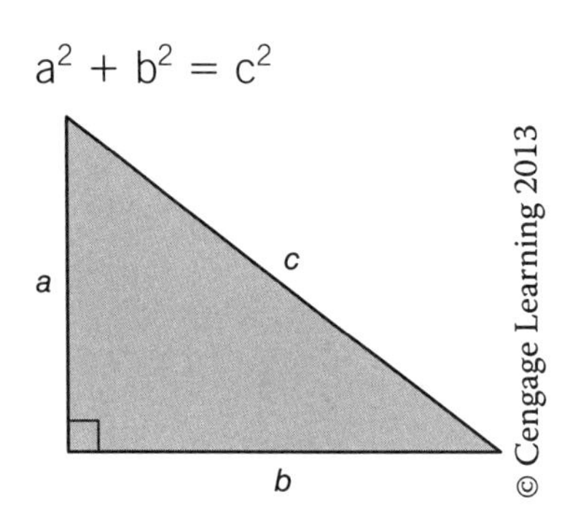

EXAMPLE 1: What is the length of the hypotenuse of the following triangle?

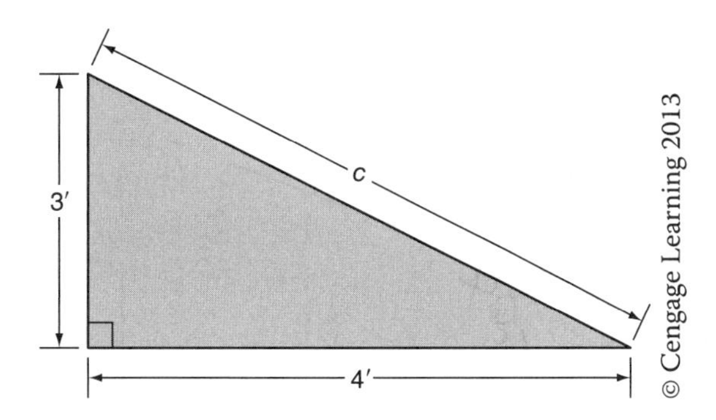

STEP 1: Set up the problem using the Pythagorean Theorem.

$a^2 + b^2 = c^2$

$3^2 + 4^2 = c^2$

STEP 2: Solve.

$9 + 16 = c^2$

$25 = c^2$

$c = \sqrt{25}$

$c = 5'$

ANSWER: The hypotenuse of this triangle measures 5 feet.

Understanding powers and roots enables masons to use the the Pythagorean Theorem to lay out and check the square alignment of two adjacent walls.

Practical Problems

Use a calculator to solve the following problems. Round off the answers to the nearest hundredth (two places to the right of the decimal). Give the equivalent measurement in feet, inches, and fractions of inches (sixteenths, eighths, quarters, or halves).

1. $\sqrt{\dfrac{12^2}{4}}$ ____________

2. $\sqrt{\dfrac{8^3}{2}}$ ____________

3. Using the formula $H = \sqrt{(6'')^2 + (8'')^2}$, find the value of H, the hypotenuse. ____________

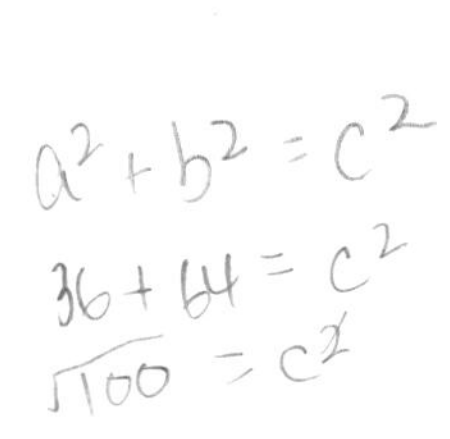
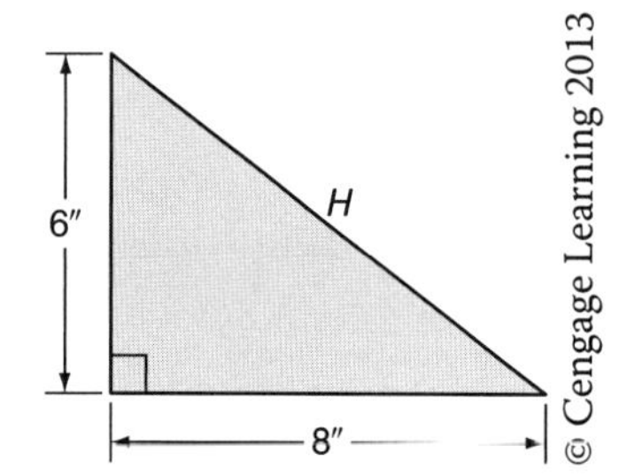

4. Using the following formula, find altitude A of this triangle: ____________

$$A = \sqrt{(10'2'')^2 - (8'0'')^2}$$

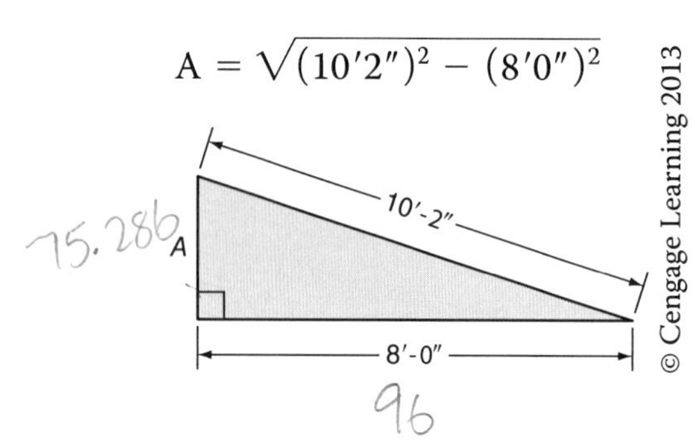

$75.286^2 + 96^2 = c^2$

$5668 + 9216 = c^2$

5. This set of steps has a rise of 8 feet 6 inches and a run of 6 feet 3 inches. What is the distance between point A and point B? Use the following formula and round off the answer to the nearest hundredth foot. ________

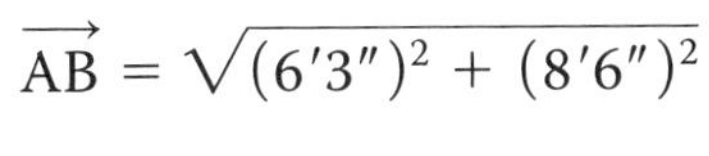

$$\overrightarrow{AB} = \sqrt{(6'3'')^2 + (8'6'')^2}$$

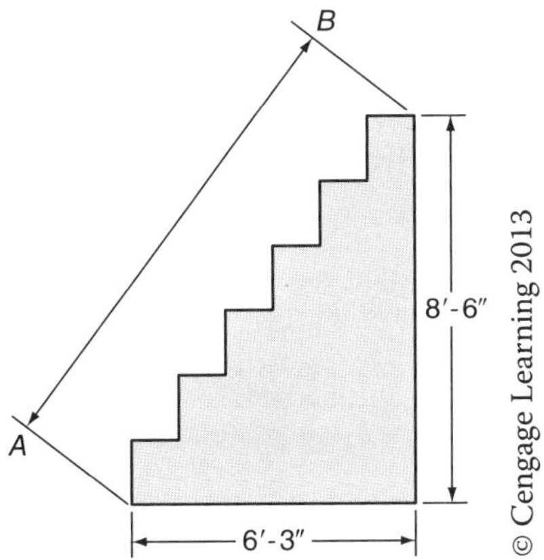

6. What should be the length of either of the two components of a scaffold cross-brace if the scaffold end frames are spaced 7 feet apart and the distance between the ends of each cross-brace is 3 feet 4 inches? ________

Answer ________′ = ________′ ________″

Round off the answer to the nearest hundredth inch. Convert the decimal format of feet into feet and inches.

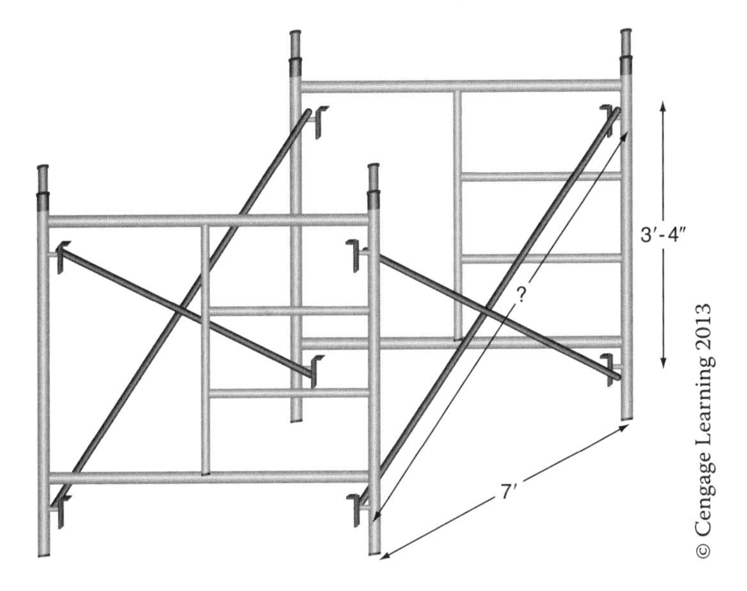

7. Batter boards, boards secured to stakes for attaching string lines, are relied upon to lay out footings and foundation walls. String lines are fastened to batter boards, a combination of horizontal rails and stakes, to lay out a foundation 42′ 8″ × 22′. What should be the length of either diagonal measurement if the string lines are aligned perpendicular at each corner? __________

   Answer ________′ = ________′ ________″

   Round off the answer to the nearest hundredth inch. Convert the decimal format of feet into feet and inches.

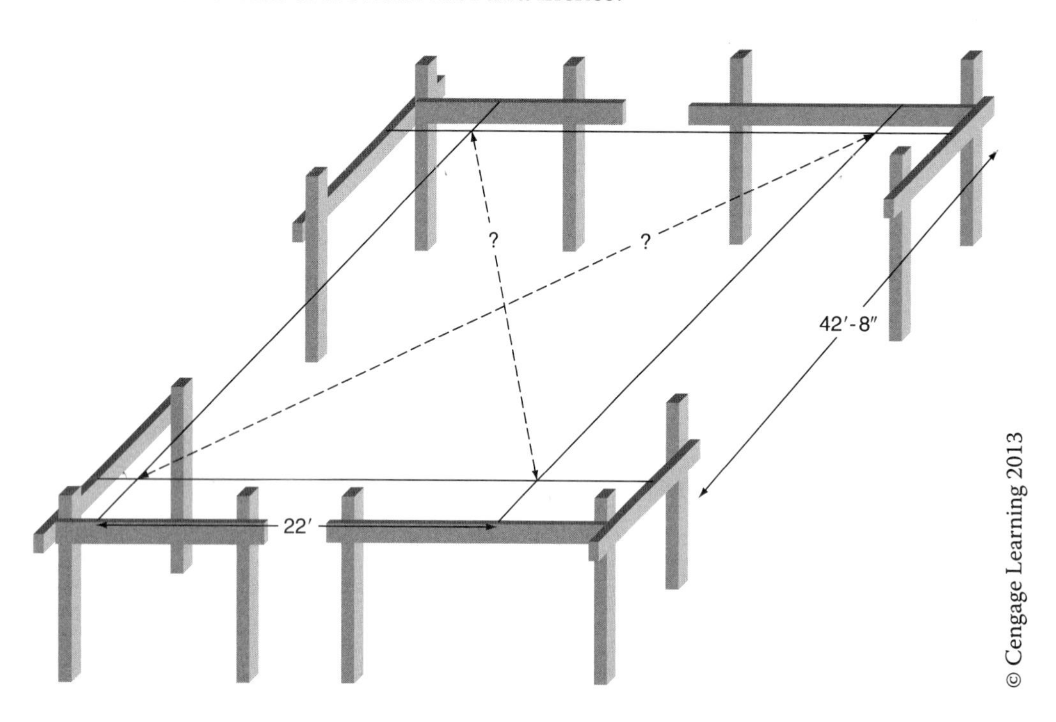

SECTION 6

Measure

UNIT 24

Metric Measure

Basic Principles of Metric Measure

The metric system is a set of measurements developed in the 1790s, primarily by the French, in an effort to create a uniform system of measurement. Thomas Jefferson was an early proponent of the use of the system, and the United States was the first country to develop its coinage based on metrics. Our dollar is divided into 100 cents. Prior to metrics, English measure consisted of a multitude of measurements, some of which we are familiar with; others have meanings that are antiquated.

Although the United States still uses a mixture of English and metrics, most other countries in the world use only the metric system. The *meter* is the standard unit of measurement of length in the metric system. It is several inches longer than the English measure.

The meter is divided into 100 small units called *centimeters,* or 1,000 smaller parts called *millimeters.* Similarly, 1,000 meters is a *kilometer.*

English Length Measure		
12 inches (in.)	=	1 foot (ft)
3 feet (ft)	=	1 yard (yd)
1,760 yards (yd)	=	1 mile (mi)
5,280 feet (ft)	=	1 mile (mi)

Metric Length Measure		
10 millimeters (mm)	=	1 centimeter (cm)
10 centimeters (cm)	=	1 decimeter (dm)
10 decimeters (dm)	=	1 meter (m)
10 meters (m)	=	1 dekameter (dam)
10 dekameters (dam)	=	1 hectometer (hm)

English–Metric Equivalents Length Measure		
1 inch (in.)	≈	25.4 millimeters (mm)
1 inch (in.)	≈	2.54 centimeters (cm)
1 foot (ft)	≈	0.3048 meter (m)
1 yard (yd)	≈	0.9144 meter (m)
1 mile (mi)	≈	1.609 kilometers (km)
1 millimeter (mm)	≈	0.03937 inch (in.)
1 centimeter (cm)	≈	0.39370 inch (in.)
1 meter (m)	≈	3.28084 feet (ft)
1 meter (m)	≈	1.09361 yards (yd)
1 kilometer (km)	≈	0.62137 mile (mi)

Note: ≈ indicates *approximately equal.*

Other standard measures used in the metric system are *grams,* a measure of mass, and *liters,* a measure of volume.

EXAMPLE 1: An outer length of wall equals 26 feet. How many meters is this?

$26 \times 0.3048 = 7.92$

ANSWER: The wall measures ≈ 7.92 meters.

Practical Problems

1. How many centimeters are there in 1 meter? ________
2. How many centimeters are there in 275 millimeters? ________
3. How many millimeters are there in 26 centimeters? ________
4. A standard brick that is 8 inches long is how many centimeters long? ________
5. A patio with a length of 20 feet 6 inches is how many meters long? ________
6. If a wall is 4.5 meters long, how many 6 inch bricks will it take to fit the length of the wall? ________

UNIT 25

Rule or Tape Having 1⁄16 Inch Graduations

Basic Principles for Using a Rule or Tape

Tasks performed by masons frequently require reading measuring rules and tapes. Masons use a variety of measuring tools, but the more common ones are the 6 foot folding rule and steel tapes. The rules and tapes masons prefer have 1⁄16 inch graduations, meaning that a single inch is divided into 16 equal parts. Rules and tapes having graduations smaller than 1⁄16 inch are not necessary for masonry construction because size tolerances in construction seldom exceed 1⁄16 inches.

Reading a Ruler

Many 100 foot steel tapes are imprinted with 1⁄8 inch graduations, increments small enough for laying out footings and foundation walls. The following representation of 1 inch shows the correct interpretations of 1⁄8 inch graduations.

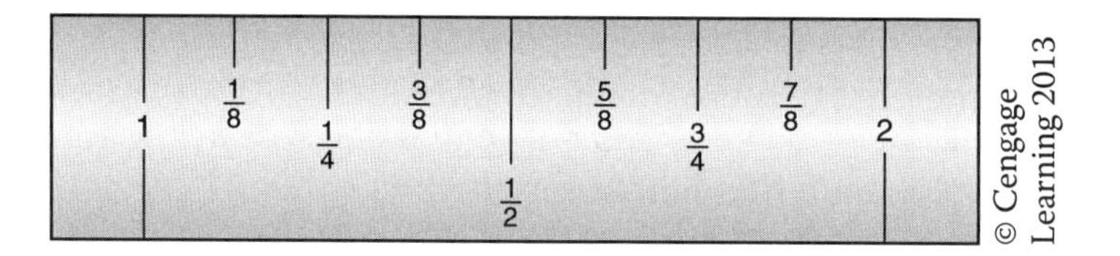

Six-foot folding rules and steel tapes shorter than 100 feet are imprinted with $\frac{1}{16}$ inch graduations, an increment sometimes needed for the layout of wall openings and other tasks. The following illustration of a rule imprinted with $\frac{1}{16}$ inch graduations is typical of most rules and tapes. A single inch is divided into 16 equal parts, making each part equivalent to $\frac{1}{16}$ inch.

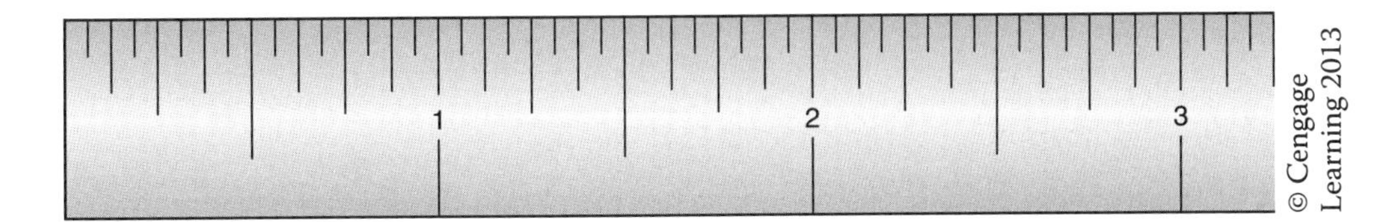

The graduation marks imprinted on rules and tapes are intended solely as dividing lines, separating equal fractional parts of an inch from one another. When learning to read a rule, remember to count the parts, that is, the fractional parts between the imprinted lines. This can be related to slicing a pizza into equal parts. Do you want the parts or the lines separating the parts? Of course, you want the parts!

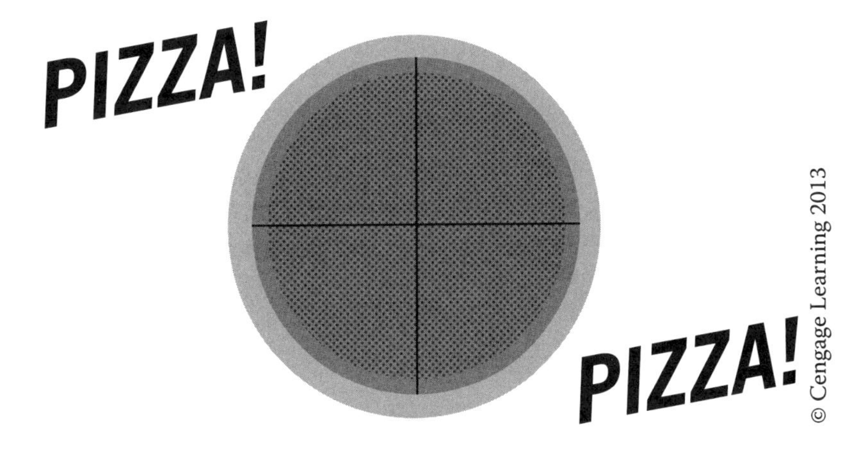

As with the pizza illustrated here, we recognize and value the *parts* and not the lines separating the parts. This pizza has been divided into four equal parts. When reading a rule, remember to count the parts rather than the lines between the parts.

The following illustrations show the correct interpretation for each of the eight 1/16 inch parts or increments of 1/2 inch.

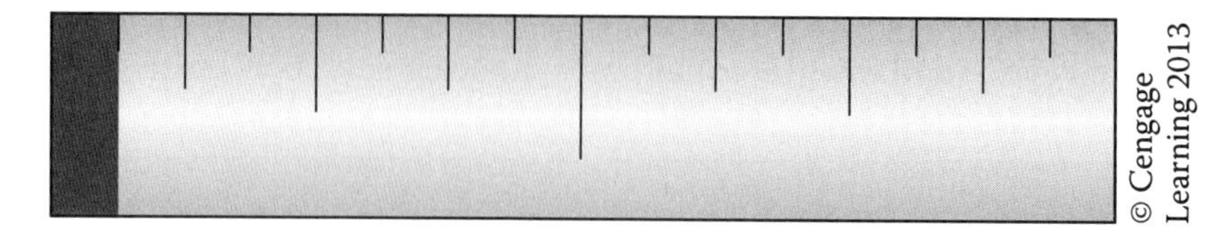

The shaded increment on the preceding rule is read as 1/16 inch.

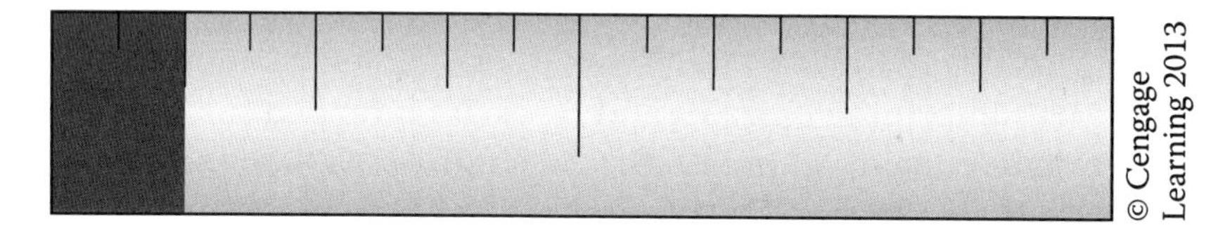

The shaded increment on the preceding rule is read as 2/16 inch, simplified as 1/8 inch.

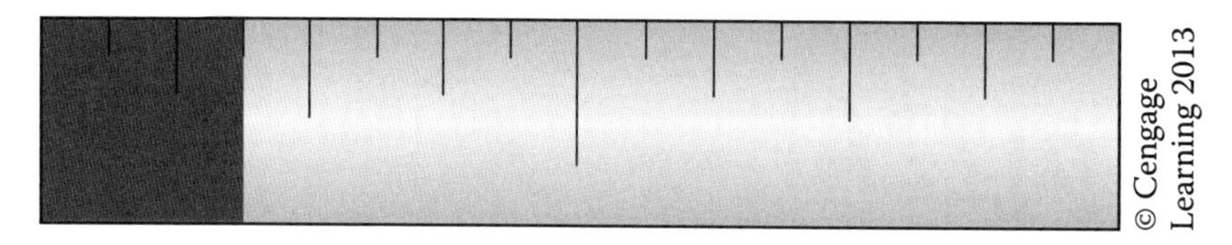

The shaded increment on the preceding rule is read as 3/16 inch.

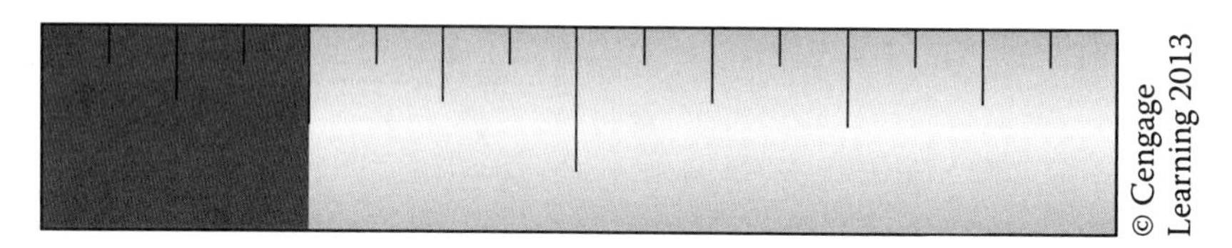

The shaded increment on the preceding rule is read as 4/16 inch, simplified as 1/4 inch.

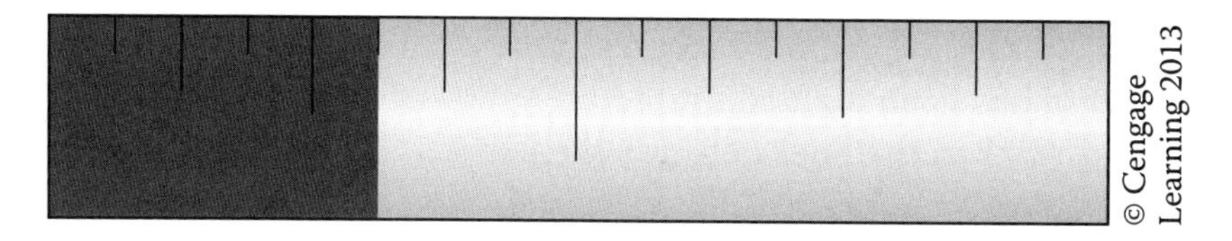

The shaded increment on the preceding rule is read as 5/16 inch.

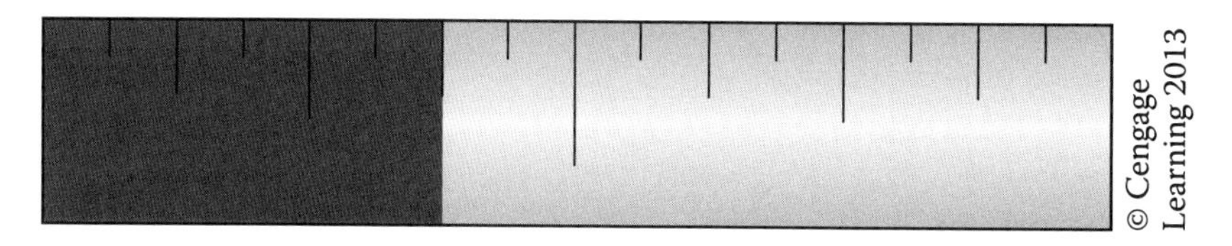

The shaded increment on the preceding rule is read as 6/16 inch, simplified as 3/8 inch.

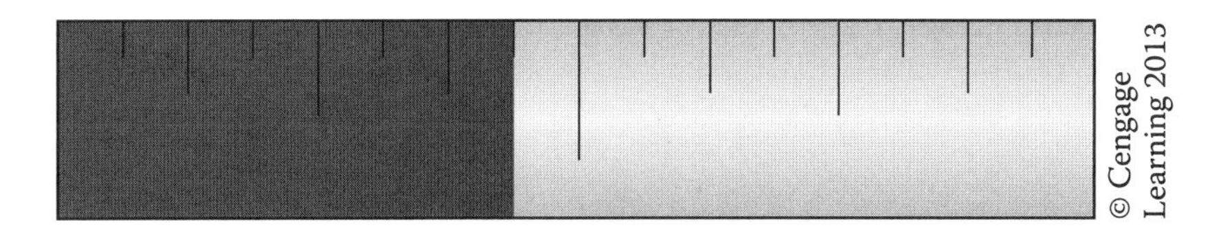

The shaded increment on the preceding rule is read as 7⁄16 inch.

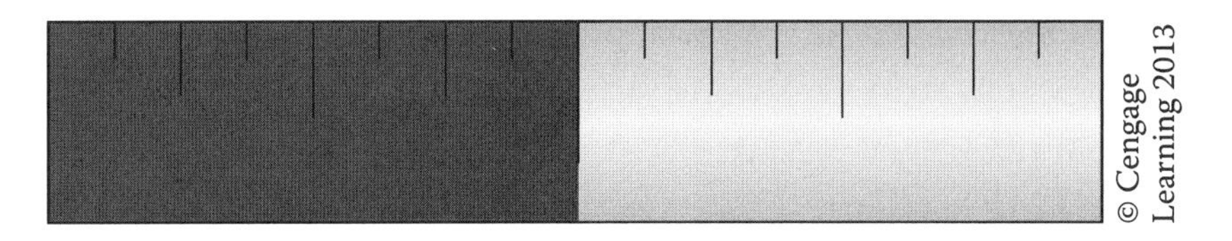

The shaded increment on the preceding rule is read as 8⁄16 inch, simplified as ½ inch.

EXAMPLE 1: The ruler below represents 1 inch. What is the correct reading for the shaded portion?

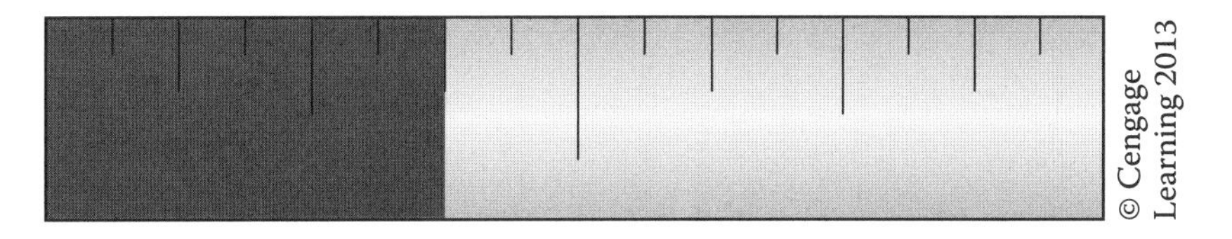

STEP 1: Count the shaded parts (not the lines).

The 1 inch increment is divided into 16 equal parts. There are six shaded parts, so this shaded portion is equal to 6⁄16 inch.

STEP 2: Simplify if possible.

$6/16'' = 3/8''$

ANSWER: The shaded portion of the ruler is read as 3⁄8 inch.

Practical Problems

Refer to the following illustration to solve problems 1–3.

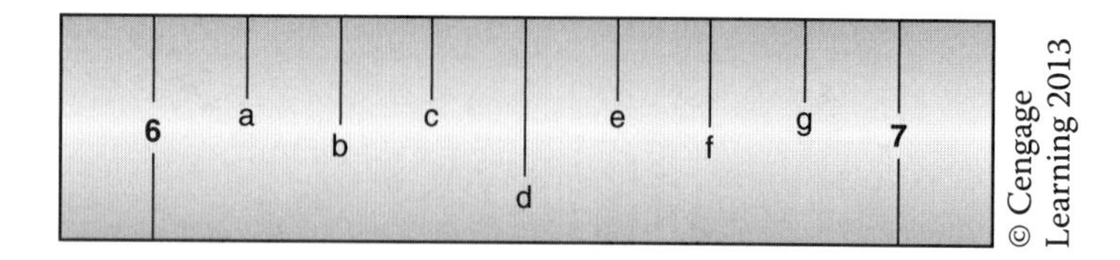

1. Into how many equal parts is a single inch divided? ______________

2. Each of the equal parts is equivalent to what fractional part of an inch? ______________

3. Provide the correct reading for each of the readings on the ruler. Simplify the fractions using the lowest common denominator.

   a = ______________"
   b = ______________"
   c = ______________"
   d = ______________"
   e = ______________"
   f = ______________"
   g = ______________"

4. Provide the correct readings on the following rule. Simplify fractions, using the lowest common denominator.

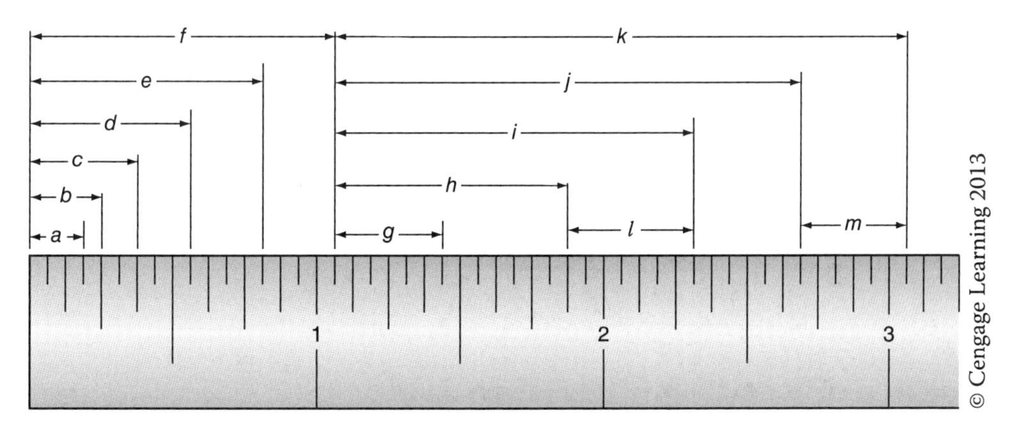

a = ________"

b = ________"

c = ________"

d = ________"

e = ________"

f = ________"

g = ________"

h = ________"

i = ________"

j = ________"

k = ________"

l = ________"

m = ________"

UNIT 26

Reading a Leveling Rod Having ⅛ Inch Graduations

Basic Principles of Reading a Leveling Rod

Leveling rods are multisection, telescoping aluminum, fiberglass, or wood tubular rods with either ⅛ inch graduations or 1/10 inch graduations imprinted on the face. Builders rely upon leveling rods and leveling instruments primarily to determine grades and elevations or to lay out and level foundation walls. Provided the leveling instrument is used properly and is not moved, the difference between leveling rod readings at different locations represents the difference in elevation or height between the surfaces where the base of the leveling rod is placed.

An understanding of the operation of a leveling instrument makes interpreting leveling rod readings simple. A leveling instrument is secured atop a tripod and leveled. Leveling the instrument assures its accuracy in all 360° of a horizontal plane, a plane aligned with the horizontal cross-hair of an optical instrument or the light beam of a laser level. Sighting through an optical instrument and recording the reading aligned with the cross-hair on the leveling rod allows one to compare readings taken at different locations. A laser level relies upon a small battery-powered motor to spin a mirror on its axis and in so doing projects a single laser beam of light continuously in 360° horizontal and level planes. A battery-operated detector mounted on a leveling rod, capable of sliding up or down the rod, detects the laser beam, giving both visual and audible signals when it is in line with the beam. The position of the sensor on the leveling rod is recorded for a comparison of readings taken at different locations.

EXAMPLE 1: In the following illustration, a laser level projects a laser beam, a single beam of light projected by a mirror continuously in a full circle. The leveling rod with detector is positioned in two different locations. What is the difference in elevation at the two locations?

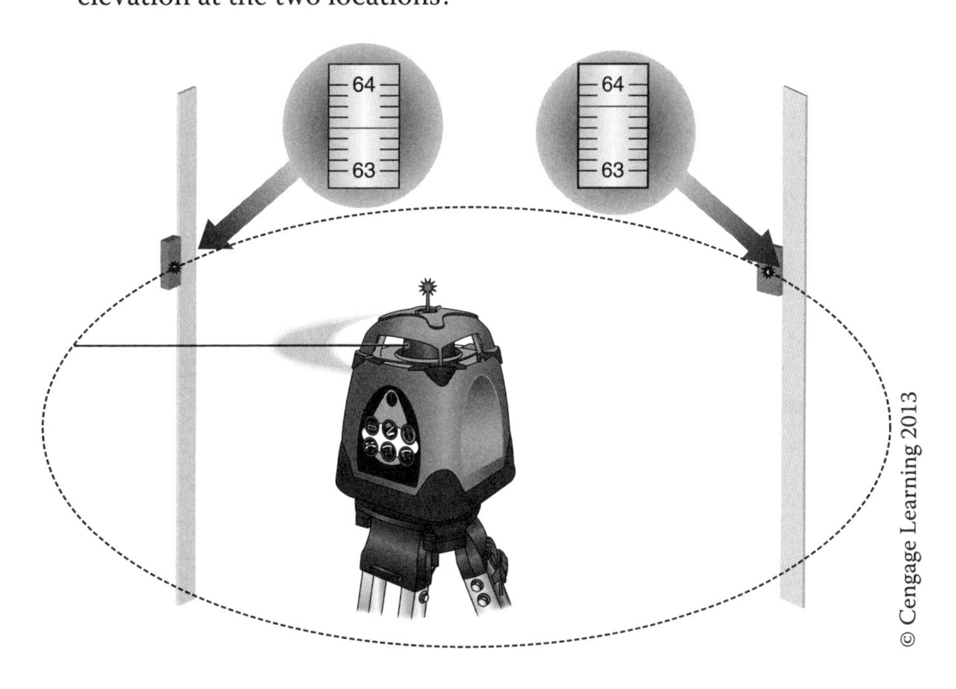

STEP 1: Read the two measurements.

The reading to the left of the instrument, where the laser beam and detector are aligned level, is 63½ inches. The reading to the right of the instrument is detected on the leveling rod at 63¾ inches.

STEP 2: Find the difference.

$63\frac{3}{4}'' - 63\frac{1}{2}'' = 63\frac{3}{4}'' - 63\frac{2}{4}'' = \frac{1}{4}''$

ANSWER: There is a difference of ¼ inch between the two readings, indicating a ¼ inch difference in elevation between the two surfaces where the base of the leveling rod is positioned.

The readings on the leveling rod increase, that is, larger measurements are recorded, as the base of the leveling rod is placed on lower surfaces. Smaller readings indicate the base of the leveling rod is placed on higher surfaces. In the preceding example, the rod reading of 63¾″ indicates that the base of the leveling rod is ¼ inch lower than where the base of the rod is placed for a reading of 63½″. Larger numbers correlate with lower surfaces on which the base of the rod is placed.

Masons use rotary laser levels or optical levels to ensure level footings and foundation walls. With proper set-up of the leveling instrument, identical readings on the leveling rod as it is moved from corner to corner confirm that the corners are level with one another.

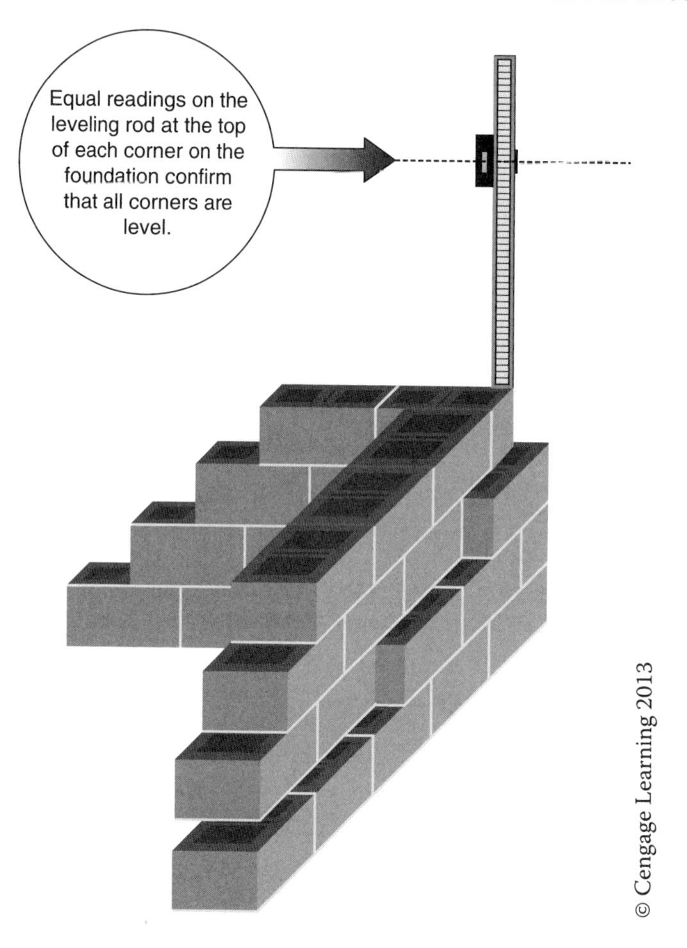

The following illustration shows how to read a leveling rod having ⅛ inch graduations. Unlike a rule or tape, the imprinted black markings serve as measured increments. Each white space is ⅛ inch wide and each black marking is ⅛ inch wide. Studying the exploded view, one can learn to accurately read a leveling rod.

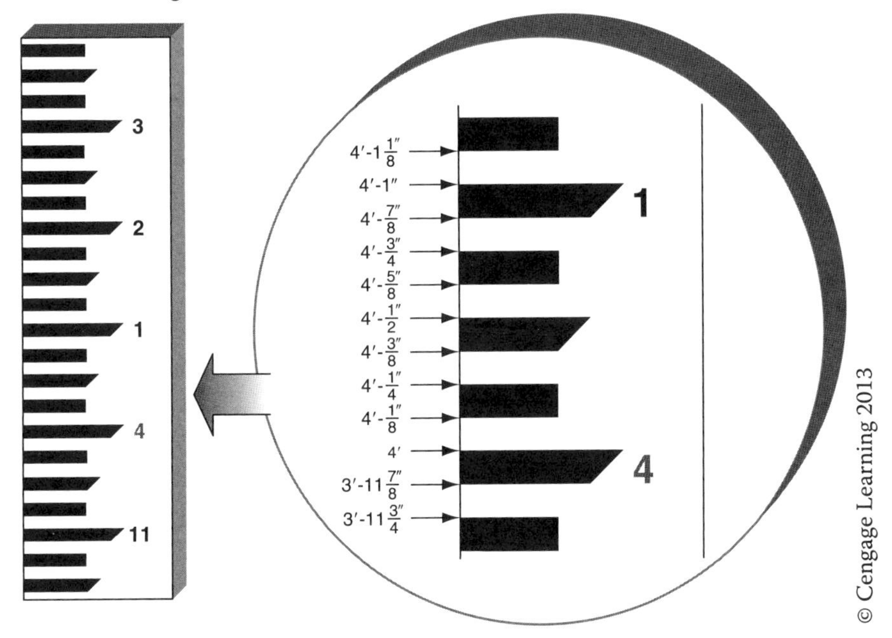

Practical Problems

Fill in the blanks for the correct readings on the following leveling rod.

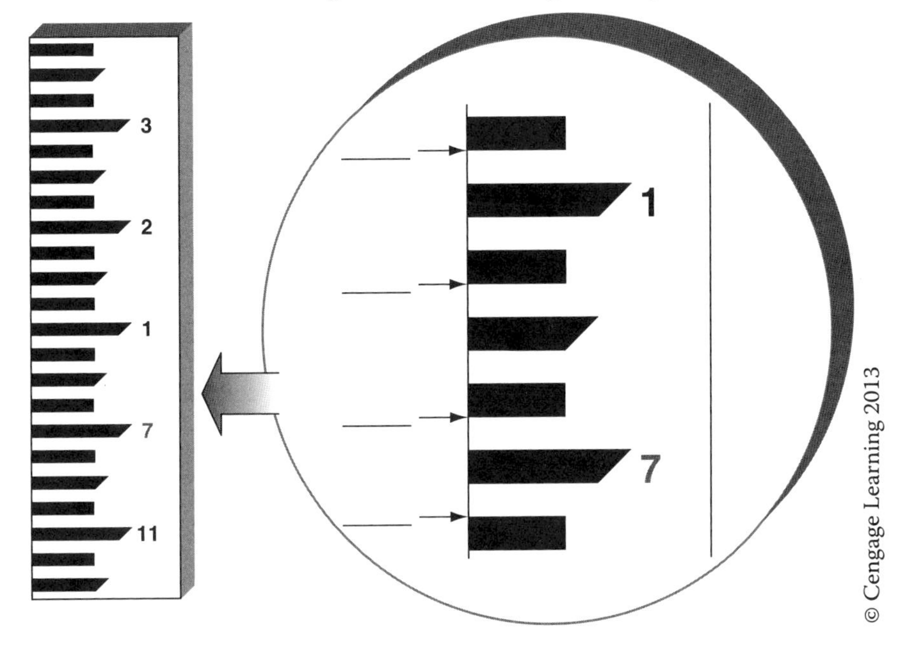

A leveling rod is used to check the finished height of a foundation wall at each of its four corners. The readings taken at the top of each corner are:

- ❑ Corner #1 – reading 14′3⁄8″
- ❑ Corner #2 – reading 13′117⁄8″
- ❑ Corner #3 – reading 14′1⁄4″
- ❑ Corner #4 – reading 14′1⁄2″

1. What is the difference in height between the lowest and highest corner? __________
2. Which of the four corners is highest? __________
3. Which of the four corners is lowest? __________
4. What is the difference in height between the highest and the lowest corner? __________

   a. Corners #1 and #2? __________

   b. Corners #2 and #3? __________

   c. Corners #3 and #4? __________

   d. Corners # 4 and #1? __________

SECTION 7

Computing Geometric Measure, Area, Volume, Mass, and Force

UNIT 27

Area of Rectangles, Triangles, and Circles

Basic Principles of Area of Rectangles, Triangles, and Circles

Area is a quantity expressed in terms of square units, identifying the size of a two-dimensional surface. The area of concrete slabs, wall openings, and walls is of particular interest to masons when estimating materials. Area is expressed as *square* units because it is the product of two dimensions, either length multiplied by width or length multiplied by height. The units typically used in construction to express area are *square inches* and *square feet.*

The product of the length of an object multiplied by its width represents an object's horizontal surface area. The areas of floor spaces, roofs, and land are sometimes referred to as their foot-print sizes, because *area* is an expression of the space enclosed or covered. The product of the length of an object multiplied by its height represents the surface area of walls and other vertical surfaces.

Area of a Square

Square (definition): a rectangle having all four sides of equal length and four equal angles, each 90° or right angles—a regular quadrilateral.

area (A) of a square, $A = s^2$

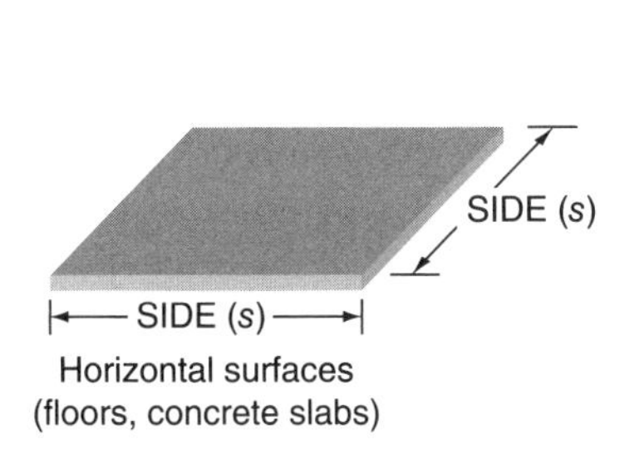

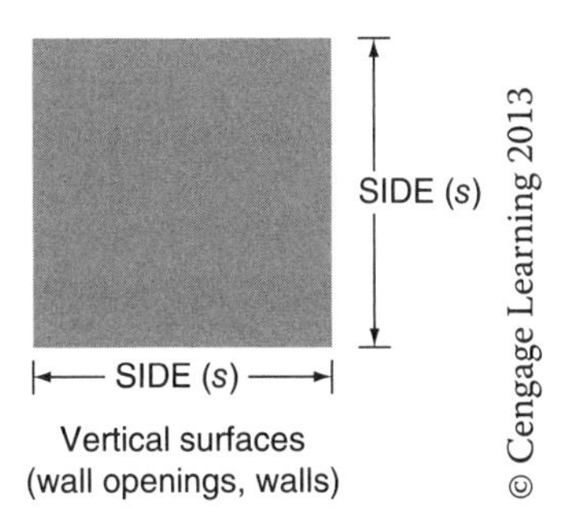

EXAMPLE 1 What is the area of a patio 6 foot long and 6 foot wide?

STEP 1: Set up the formula $A = lw$.

$A = (6')(6')$

STEP 2: Multiply.

$6 \times 6 = 36$

ANSWER: The area of this patio is 36 square feet.

Area of a Rectangle

Rectangle (definition): a quadrilateral with opposite sides of equal lengths and with four 90° or right angles.

EXAMPLE 1 What is the area of a wall whose length is 14 feet 6 inches and height is 9 feet 4 inches?

area (A) of a rectangle
$A = lw$

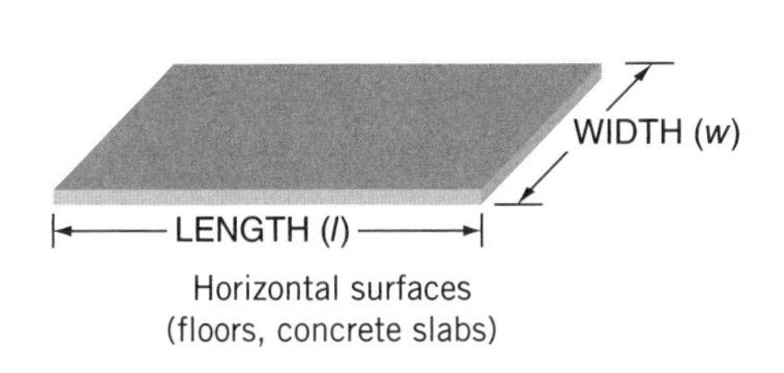

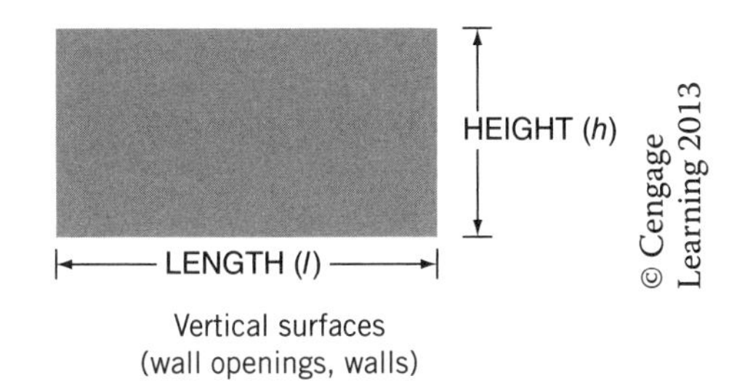

STEP 1: Set up the formula A = lw.

A = (14.5′)(9.33′)

STEP 2: Multiply.

$14.5 \times 9.33 = 135.285$

ANSWER: The area of this wall is 135.29 square feet.

Area of a Triangle

Triangle (definition): a figure having three angles adding up to 180° and three straight sides; some or all of the angles or lengths of the sides may or may not be equal.

area (A) of a triangle

$A = \frac{1}{2}bh$

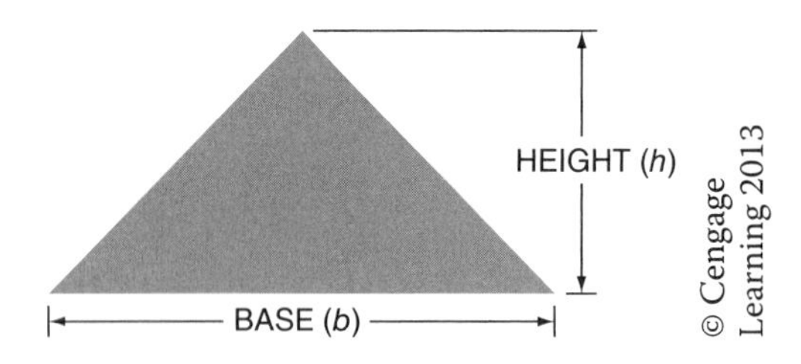

EXAMPLE 1 What is the area of this triangle?

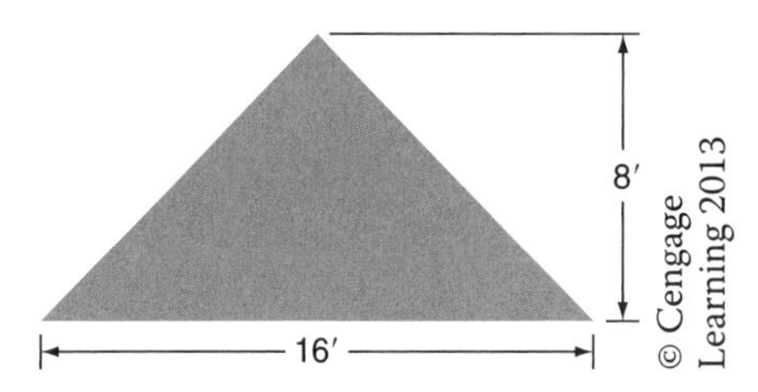

STEP 1: Set up the formula $A = ½bh$.

$A = ½(16 \times 8)$

STEP 2: Multiply.

$A = ½(128)$

ANSWER: The area of this triangle is 64 square feet.

Area of a Circle

Circle (definition): a round-shaped figure or object whose perimeter line or circumference is always an equal distance, known as its *radius*, from its center point.

area (A) of a circle $= \pi r^2 = 3.14 \times \text{radius}^2$

$A = \pi r^2$

$A = (3.14)r^2$

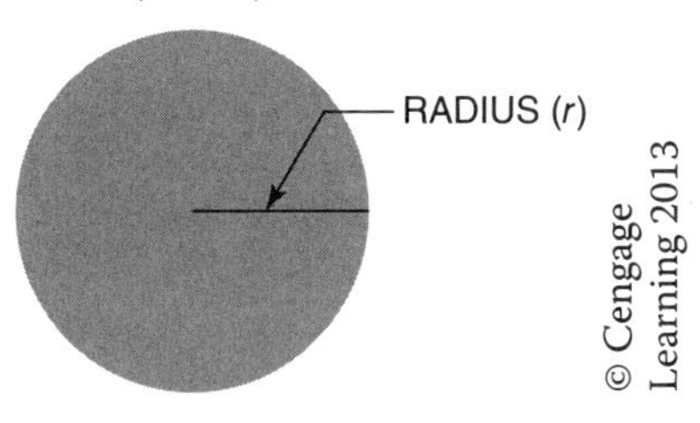

EXAMPLE 1 What is the area of a circle having a radius of 5 feet?

STEP 1: Set up the formula $A = \pi r^2$

$A = 3.14 \times 5^2$

STEP 2: Multiply.

$A = 3.14 \times 25 = 78.5$

ANSWER: The area of this circle is 78.5 square feet.

Practical Problems

Find the area of each of the following figures.

1.

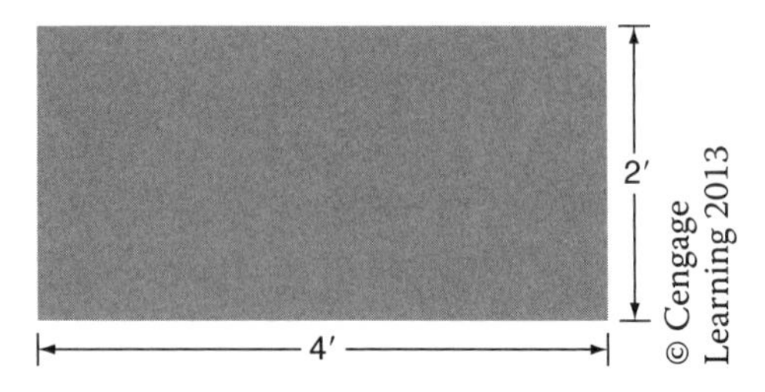

The area of the rectangle is ________ square feet. ________________

2.

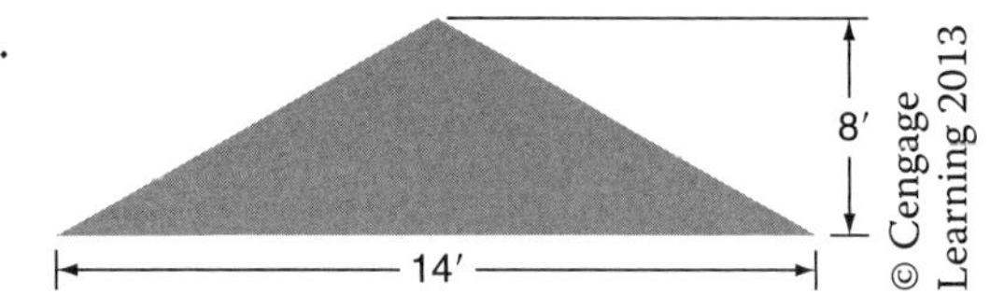

The area of the triangle is ________ square feet. ________________

3.

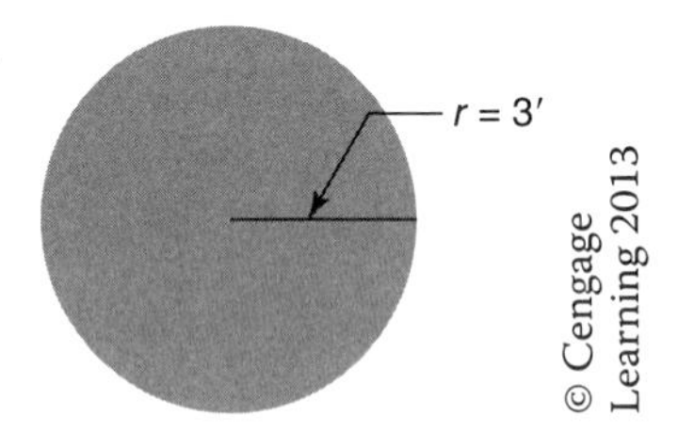

The area of the circle is ________ square feet. ________________

The following illustration depicts one section of a tubular welded scaffold. Refer to it to solve problems 4 and 5.

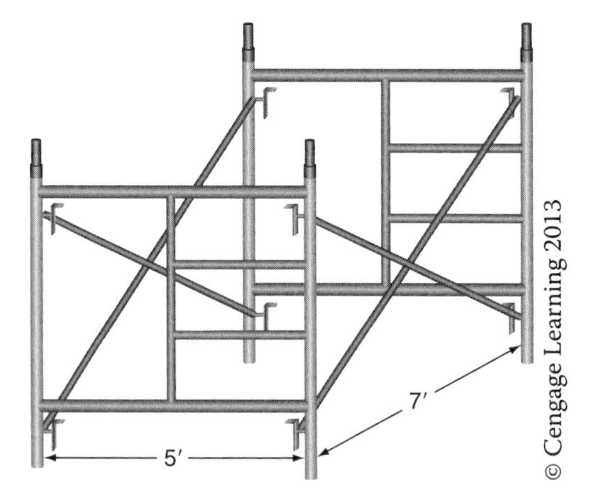

4. While scaffold planks may be required to extend at least 6 inches beyond their end supports (scaffold end frames), the maximum weight that any one scaffold section can support is based upon the span between the end supports and the overall width of the planked section. What is the maximum total area of planking that can be supported between the end frames of this one section of scaffold? ____________

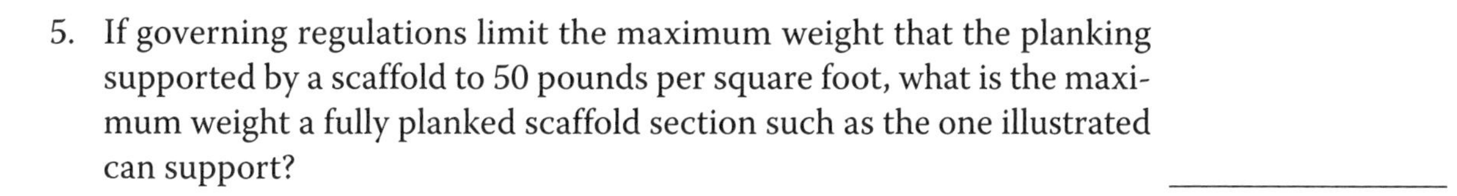

5. If governing regulations limit the maximum weight that the planking supported by a scaffold to 50 pounds per square foot, what is the maximum weight a fully planked scaffold section such as the one illustrated can support? ____________

Refer to the following illustration to solve problems 6–8. A hollow concrete masonry unit (CMU) or block is one in which the cross-sectional area of the cells is more than 25% of the overall cross-sectional area of the unit. A solid CMU is defined as either having cells with a combined cross-sectional area of up to 25% of the overall cross-sectional area of the unit.

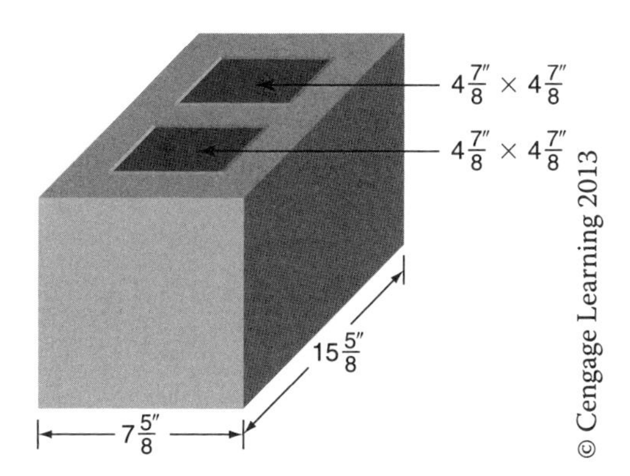

6. What is the overall cross-sectional area of the top or bottom side of the block? Give answer to the nearest thousandth inch (three decimal places). ____________

7. What is the cross-sectional area of one open block cell? Give your answer to the nearest thousandth inch (three decimal places). ____________

8. What percentage of the overall cross-sectional area of the top or bottom side of the block is the combined area of two open block cells? ____________

Refer to the following illustration of a brick to solve problems 9–11. Solid brick masonry construction is defined as construction using bricks, whose net cross-sectional area parallel to its bedding area is 75% or more of its gross cross-sectional area measured in the same plane. In other words, the combined cross-sectional area of the holes known as cores is less than 25% of the overall cross-sectional area of the top or bottom side of the brick.

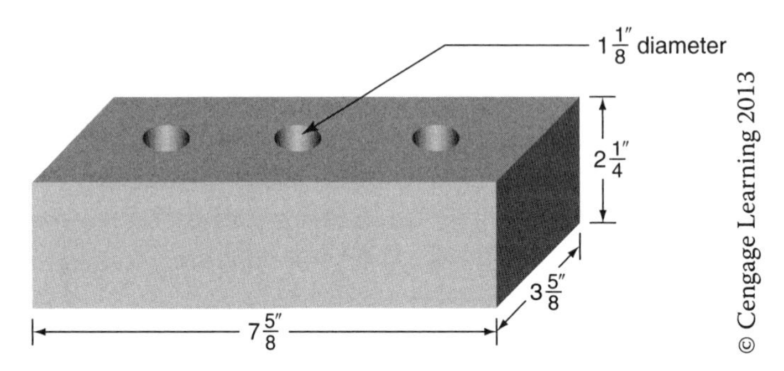

9. What is the overall cross-sectional area of the top or bottom side of the brick? Give answer to the nearest thousandth inch (three decimal places). __________

10. What is the cross-sectional area of a single core? Give answer to the nearest thousandth inch (three decimal places). __________

11. What percentage of the overall cross-sectional area of the top of the brick is the combined area of three cores? __________

Refer to the following two illustrations of foundation layouts to solve problems 12–16.

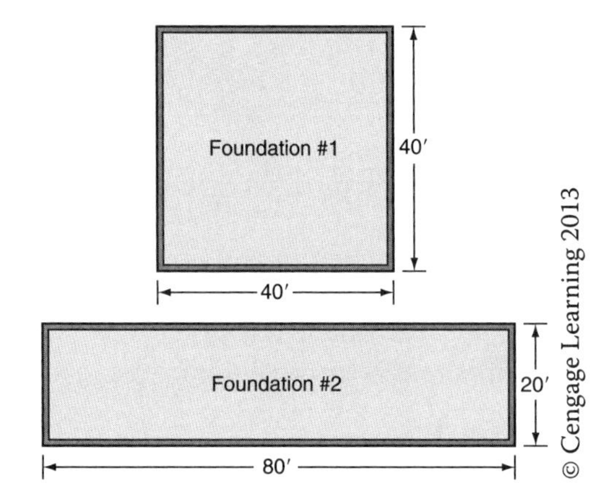

12. What is the area of foundation #1? ________ square feet __________

13. What is the area of foundation #2? ________ square feet ________

14. What is the perimeter length of foundation #1? ________ linear feet ________

15. What is the perimeter length of foundation #2? ________ linear feet ________

16. What percentage is the perimeter length of foundation #1 compared to that of foundation #2? ________% ________

17. Assume regulations require the outer hearth for fireplace openings less than 6 square feet to extend at least 16 inches in front of and 8 inches beyond each side of the opening. Assume openings that are 6 square feet and larger require extensions at least 20 inches in front of and 12 inches beyond each side of the opening. What is the minimum extension for the outer hearth of a fireplace opening having a width of 36 inches and height of 29 inches? ________

    A minimum of ________inches in front of and ________inches beyond each side.

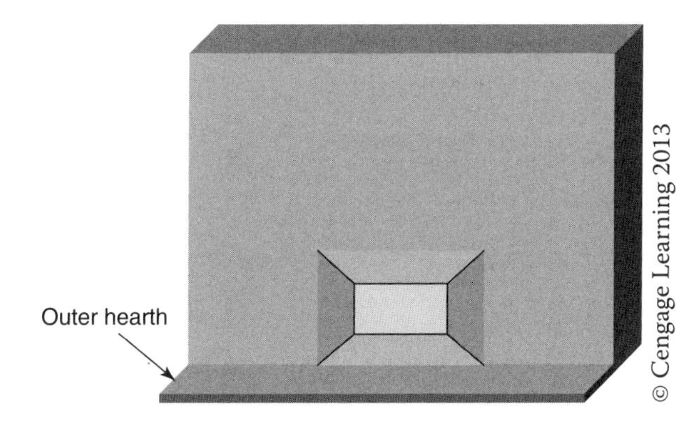

18. What is the cross-sectional (opening) area of an 8" × 12" chimney flue liner assuming the actual inside dimensions, from which the area is calculated, measure 11" × 7"? __________

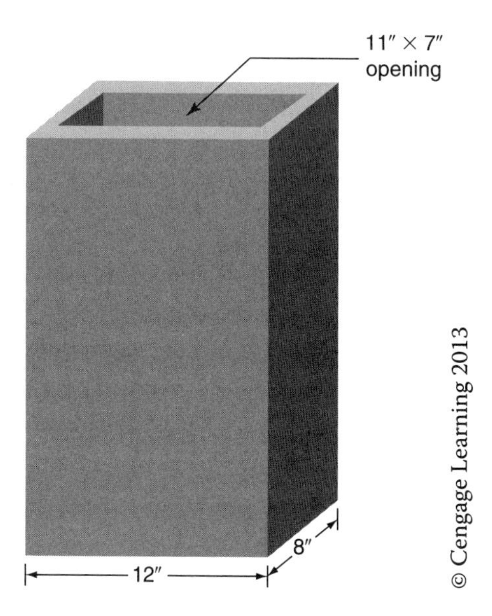

19. What is the cross-sectional opening area of a round stovepipe 8 inches in diameter, illustrated below? __________

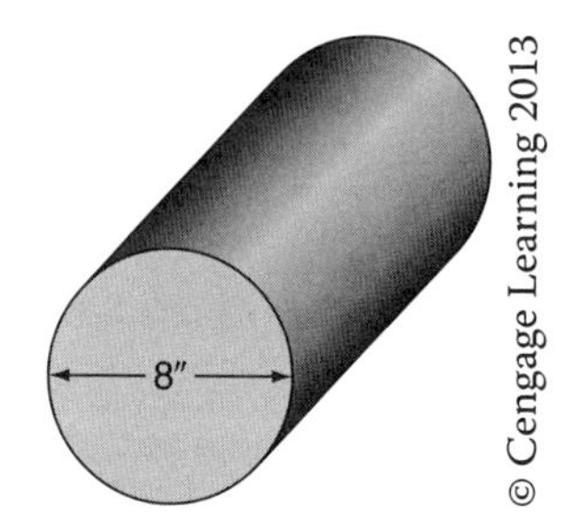

20. If regulations require that the cross-sectional area of a flue lining be equal to or larger than the cross-sectional area of a stovepipe connecting an appliance to a flue liner, can an 8 inch-round stovepipe be connected to an 8" × 8" chimney flue liner having inside measurements of 7" × 7" and meet this regulation? __________

    a. What is the cross-sectional area of the opening of this flue liner? __________

    b. What is the cross-sectional area of the 8 inch stovepipe? __________

UNIT 28

Volumes of Cubes, Rectangular Prisms, and Cylinders

Basic Principles of Volumes of Cubes, Rectangular Prisms, and Cylinders

Volume is a quantity defining the amount of space a substance (solid, liquid, or gas) occupies or an object contains. Volume is expressed as *cubic* units because it is the product of three dimensions: length, width, and height. The units typically used in construction to express volume are *cubic inches, cubic feet,* and *cubic yards.* The product of an object's surface area multiplied by its height or depth quantifies its volume. Masons need to calculate volume when estimating quantities of concrete for foundation footings, concrete floor slabs, patios, and walkways.

Volume of a Cube

Cube (definition): a three-dimensional solid object bounded by six square faces.

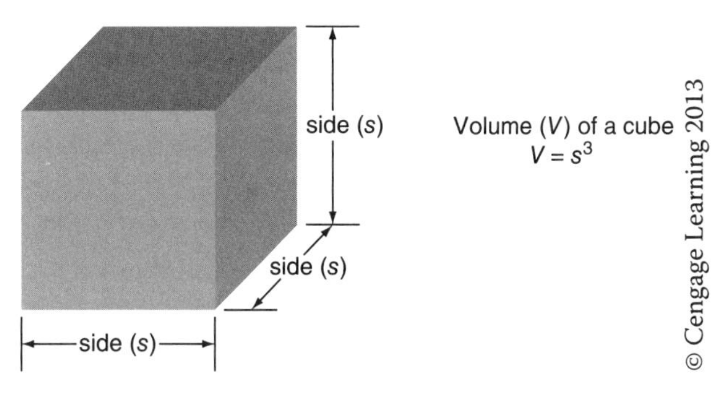

EXAMPLE 1 What is the volume of a square room with a wall height of 12 feet?

STEP 1: Set up the formula $V = s^3$

$V = 12^3$

STEP 2: Multiply.

$V = 12 \times 12 \times 12$

ANSWER: The volume of the room is 1,728 cubic feet.

Volume of a Rectangular Prism

Rectangular prism (definition): a three-dimensional solid object bound by six rectangular faces.

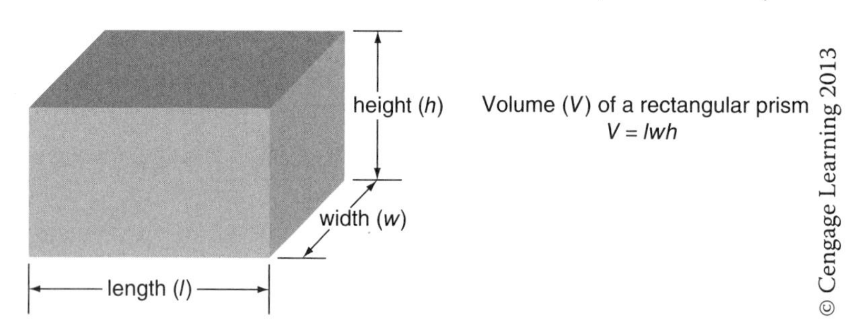

EXAMPLE 1 What is the volume of a room measuring 20' long × 16' wide × 9' tall?

STEP 1: Set up the formula $V = lwh$.

$V = (20)(16)(9)$

STEP 2: Multiply.

$V = (20 \times 16) \times 9 = 320 \times 9 = 2{,}880$

ANSWER: The volume of the room is 2,880 cubic feet.

Volume of a Cylinder

Cylindrical solid (definition): a three-dimensional solid object whose two opposite ends are round-shaped, formed by the points at a fixed distance from the axis or center line of the cylinder. If the ends are circles, then it is called a circular cylinder and the formula for its volume is given next.

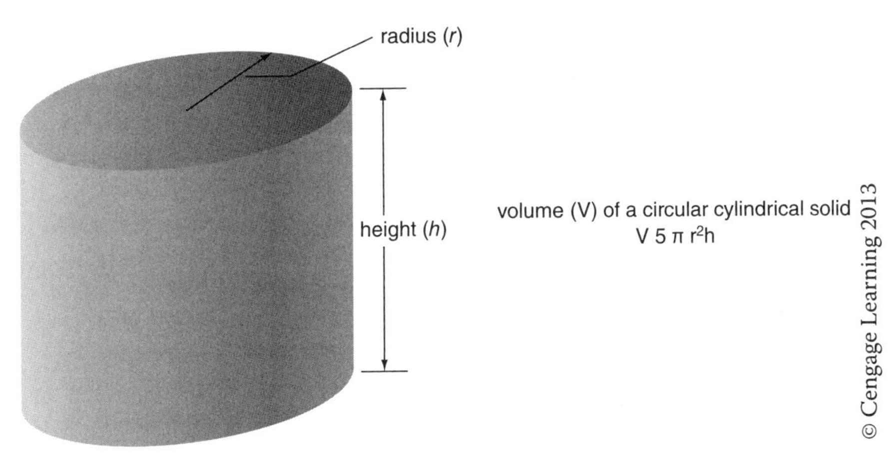

EXAMPLE 1 What is the volume of a cylindrical barrel whose diameter is 3 feet and height is 4 feet?

STEP 1: Set up the formula $V = \pi r^2 h$.

$V = (3.14)(1.5^2)(4)$

STEP 2: Multiply.

$V = 3.14 \times 2.25 \times 4$

ANSWER: The volume of this cylindrical barrel is 28.26 cubic feet.

Practical Problems

Solve the problems involving the volumes of cubes, rectangular prisms, and cylinders as related to masonry. Round off all answers to the nearest hundredth (two places to the right of the decimal).

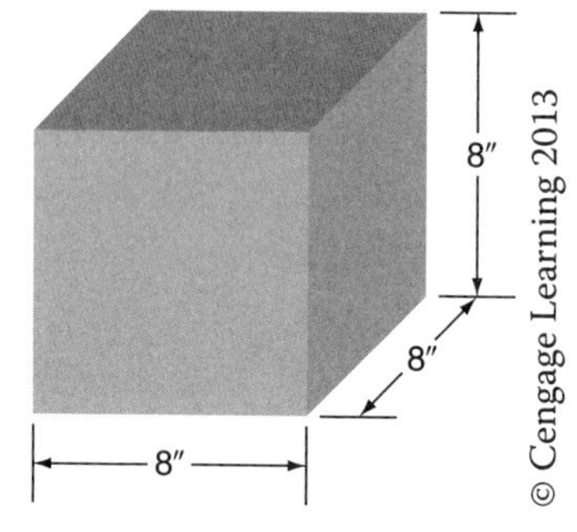

1. The volume of the cube is ________ cubic inches. ________

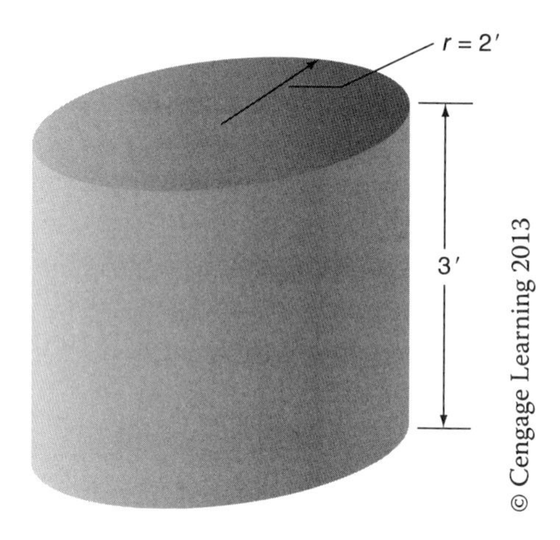

2. The volume of this cylinder is ________ cubic feet. ________

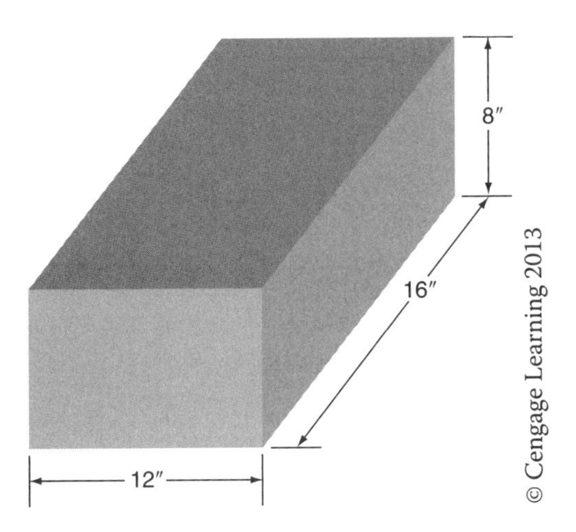

3. The volume of this rectangular prism is ________ cubic inches. ________

4. A cubic yard of concrete can be represented as a volume of concrete measuring 3 feet long, 3 feet wide, and 3 feet deep. A cube measuring 3 feet, the equivalent of 1 yard, on all sides, represents this volume, thus the expression *cubic yard.* Referring to the following illustration, how many cubic feet are there in 1 cubic yard of concrete? ________

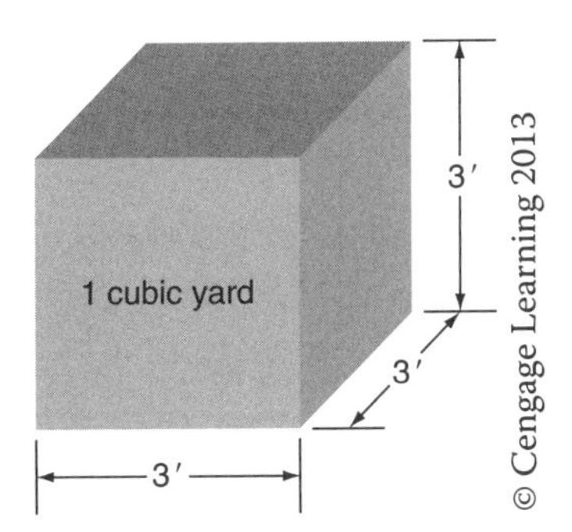

5. What is the volume of a single 12 inch concrete masonry unit (block) having a length of 15⅝ inches, width of 11⅝ inches, and height of 7⅝ inches? ________

6. Following is an illustration of a concrete footing designed to support a masonry wall, having a length of 16 feet, width of 24 inches, and thickness of 8 inches. How many cubic feet of concrete are needed to place the footing? __________

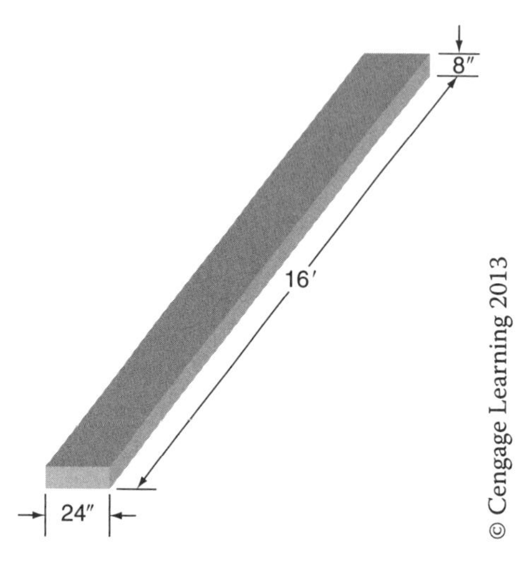

7. Convert the answer in problem 3 to cubic yards. __________

8. How many cubic yards of concrete are needed to place the 4-inch-thick concrete slab for a patio having a length of 22 feet, 4 inches and a width of 18 feet, 8 inches as illustrated next? __________

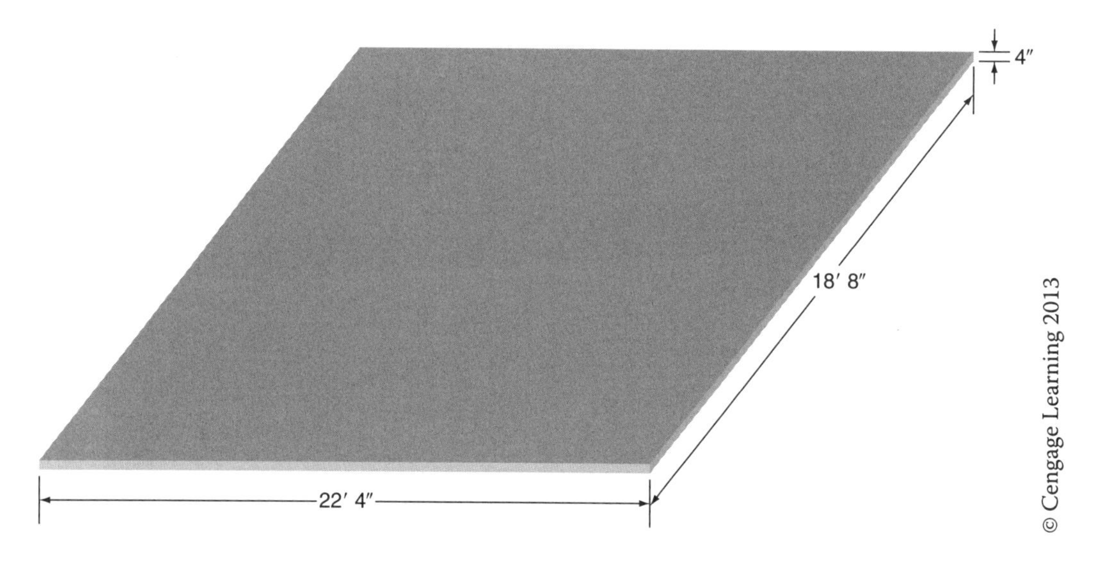

UNIT 29

Weight (Mass) Measure

Basic Principles of Weight (Mass) Measure

Weight is a measure of the force an object exerts upon its support when motionless. The weight of an object is directly proportional to a celestial body's gravitational field, the attractive force that it exerts on all objects. Because the gravitational pull of the earth's moon is just one-sixth that of the earth, a person weighing 180 pounds on earth would weigh one-sixth this weight, or 30 pounds, on the moon. Weight should not be confused with mass. *Mass* is a measure of an object's *volume.* Although one's *weight* on the moon would be one-sixth that on earth, his *mass* on the moon and the earth are equal. Mass and weight are proportional when the object is in a gravitational field.

EXAMPLE 1 If concrete weighs 145 pounds per cubic foot, what is the weight of this concrete footing?

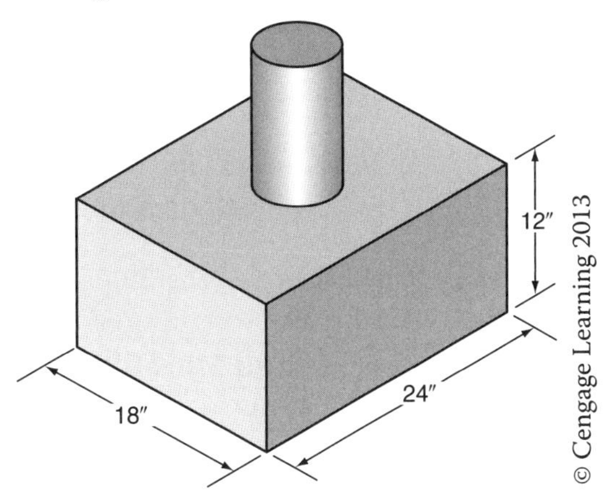

STEP 1: Set up the problem to find the volume of the footing. (Convert to feet for an answer in cubic feet.)

$V = lwh$
$V = 12" \times 18" \times 24"$
$V = 1' \times 1.5' \times 2'$

STEP 2: Multiply.

$V = 1' \times 1.5' \times 2 = 3$ cubic feet

STEP 3: Multiply by the weight of concrete per cubic feet to find the weight of the footing.

3 cubic feet × 145 pounds per cubic foot = 435 pounds

ANSWER: The concrete footing weighs 435 pounds.

Practical Problems

Solve the following problems.

1. A 12 inch I-beam weighs 31.5 pounds per linear foot. Find, to the nearest hundredth, the weight of an I-beam that is 26 feet 4 inches long. __________
2. The weight of a ¼ inch-round reinforcing bar is 0.17 pound per linear foot. What is the weight of 135 feet of this bar? __________
3. A concrete wall is 8 feet high, 32 feet long, and 12 inches thick. If concrete weighs 145 pounds per cubic foot, how much does the wall weigh? __________
4. If a single lightweight 8 inch CMU or block weighs 29 pounds and a supplier stacks 105 of them on a wood pallet weighing 35 pounds for delivery, what is the combined weight of the CMUs and the wood pallet? __________

5. If 1 cubic foot bag of masonry cement weighs 75 pounds, 1 cubic foot of damp sand weighs 82 pounds, and 1 gallon of water weighs 8.3 pounds, what is the weight of one batch of mortar made by combining a 1 cubic foot bag of masonry cement, 3 cubic feet of sand, and 4½ gallons of water? ______

6. If a standard-size modular brick weighs 4½ pounds, what is the weight of 1 cube of standard-size modular brick containing five banded groups each containing 105 bricks? ______

7. If 1 cubic foot of water weighs 62.4 pounds and contains 7.5 gallons, what is the weight of 1 gallon of water? ______

UNIT 30

Force Measure

Basic Principles of Force Measure

Force can be thought of as active power or energy, such as the force of an explosion or the force exerted by one object striking another object. Force is measured in *foot-pounds.* The term *foot-pounds* quantifies energy, the capacity of doing work. Foot-pounds is a quantity of force, or the product of the weight of an object and the distance in feet that it is permitted to travel before stopping.

Understanding force enables masons to have a working knowledge of valuable, sometimes life-saving information. For example, personal fall arrest systems (PFAS) and their components are rated according to their arresting force, which is the maximum force which a PFAS is safely capable of supporting when stopping a worker's fall in mid-air. For this application, foot-pounds of force is the product of a body's weight and the distance that it free-falls before meeting resistance. Safety regulations may require PFAS components to be capable of supporting at least 5,000 foot-pounds of force. Equipment rated capable of supporting 5,000 foot-pounds of force does not mean that the equipment is capable of supporting a 5,000-pound weight free-falling and brought to an abrupt stop. Hard hats are intended to absorb limited forces of impact from falling objects or objects encountered in head-on collisions. Forces greater than 40 pounds, a force equivalent to a 4-pound brick striking a hard hat after a fall of 10 feet, may deform or damage some hard hats.

Formula for Calculating Foot-Pounds of Force

m = the weight of an object

a = feet of travel

F (foot-pounds of force) = ma

EXAMPLE 1 How many foot pounds of force can a 195-pound worker exert on a PFAS stopping a free fall of 6 feet?

STEP 1: Set up the problem using the formula $F = ma$.

$F = (195)(6)$

STEP 2: Multiply.

$F = 195 \times 6 = 1{,}170$ foot-pounds

ANSWER: The worker exerts 1,170 foot-pounds of force.

The worker's weight is just a small fraction of the force exerted on the PFAS, a force of 1,170 foot-pounds. However, this force is well below the maximum safe rating of the components of a PFAS rated at 5,000 foot-pounds.

Note: Never equate force with a body's weight. Safety standard requirements for some safety equipment are sometimes expressed as pounds of force. Do not equate pounds of force with the weight of a body or other object.

Practical Problems

Solve each problem. Round off all answers to the nearest hundredth (two places to the right of the decimal).

1. A brick weighing 4.5 pounds free-falls 18 feet before striking the top of a worker's hard hat. How many foot-pounds of force does the brick exert on the hard hat? __________

2. A 200-pound worker slips on a sloping roof, free-falling 8 feet before contacting the safety guardrail. How many foot-pounds of force did the worker exert on the safety guardrail? __________

3. A 185-pound worker carelessly decides to jump down onto a lower scaffold level 2 feet below the level at which she is now working. How many foot-pounds of force does the worker exert on the scaffolding plank? __________

4. If a 165-pound worker free-falls 3 feet onto a top-rail, how many foot-pounds of force is exerted on the top-rail? ____________

5. Four bricks strapped together, each weighing 4.5 pounds, drop 15 feet onto a scaffolding plank. How many foot-pounds of force is exerted on the scaffold planking? ____________

6. A worker weighing 135 pounds slips and falls 5 feet onto a scaffolding plank at a lower level. How many foot-pounds of force is exerted on the scaffolding plank? ____________

SECTION 8

Formulas to Align Masonry Walls

UNIT 31

Square Columns and Piers

Basic Principles of Using Mathematics to Lay Out Square Columns and Piers

Use of parallel and perpendicular lines aids in the accurate square alignment of a variety of masonry projects, including columns, piers, pilasters, chases, and foundations. Lines are considered *parallel* to one another when they are equidistant at all points along their entire lengths.

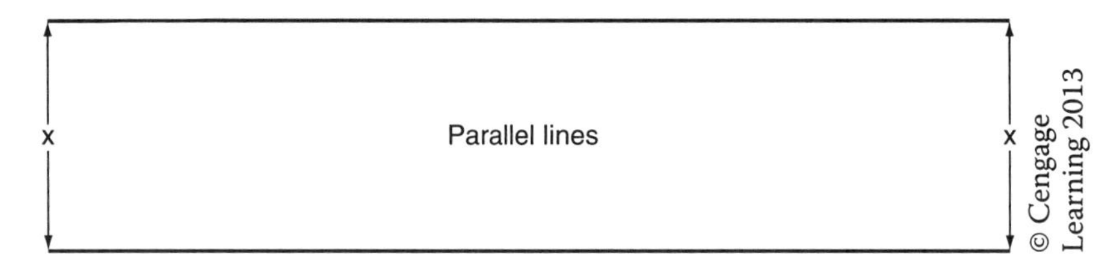

Two lines forming a right angle, or a 90° angle, are said to be *perpendicular* to one another.

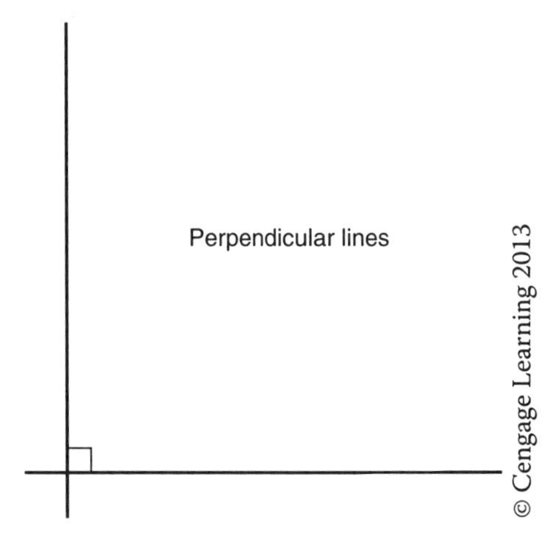

A *square* column or pier has four sides and any two adjacent sides are aligned perpendicular to one another. A column's or pier's shape is referred to as a *square* when the lengths of all four sides are equal.

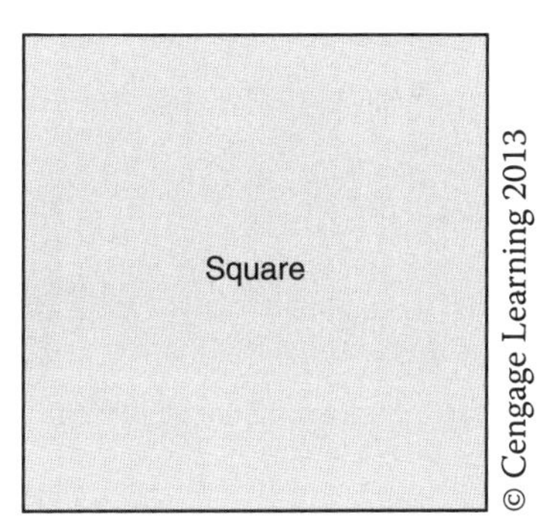

The front and back sides of a *rectangle* have equivalent lengths, while the left and right sides have a different equivalent length. All sides are perpendicular to one another.

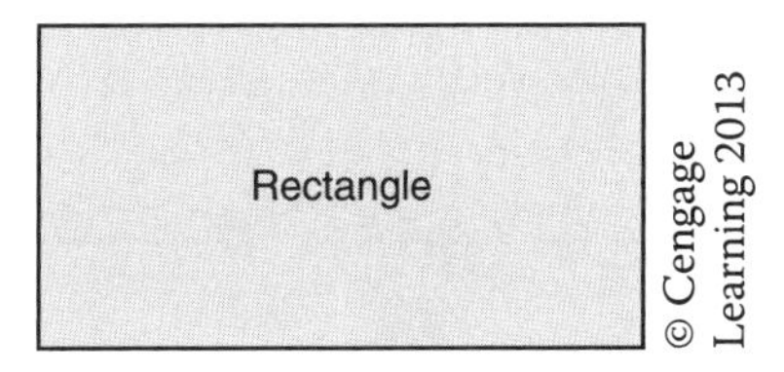

Laying Out Square Columns and Piers

The following procedures are suggested when laying out square masonry columns or piers whose base dimensions are no more than 4 feet.

STEP 1: Using a 4 foot mason's level as a straight edge, draw a straight line on the surface of the concrete slab intended to support the column or pier.

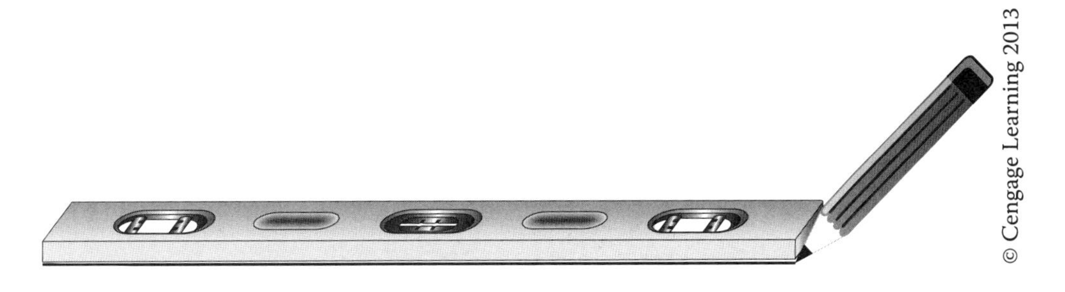

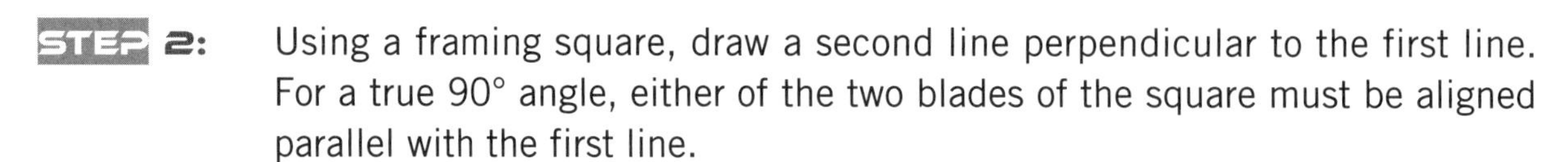

STEP 2: Using a framing square, draw a second line perpendicular to the first line. For a true 90° angle, either of the two blades of the square must be aligned parallel with the first line.

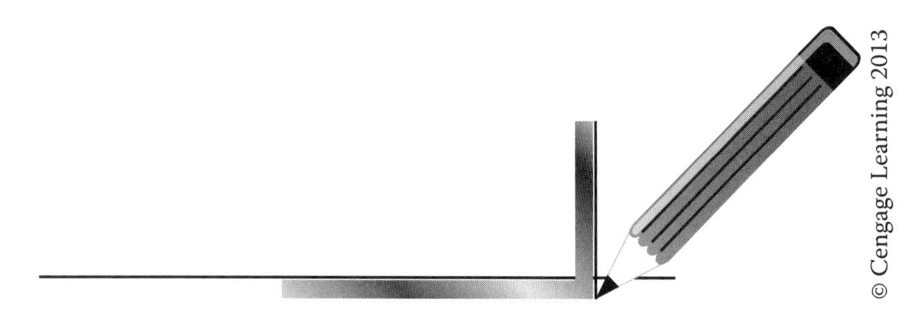

STEP 3: Using the edge of the 4 foot mason's level, lengthen this line. Continue the line further beyond its intersection with the first line. Only the extensions of the intersecting lines are easily observed once the first course of bricks is bedded in mortar and the extensions are used to properly align each of the four sides.

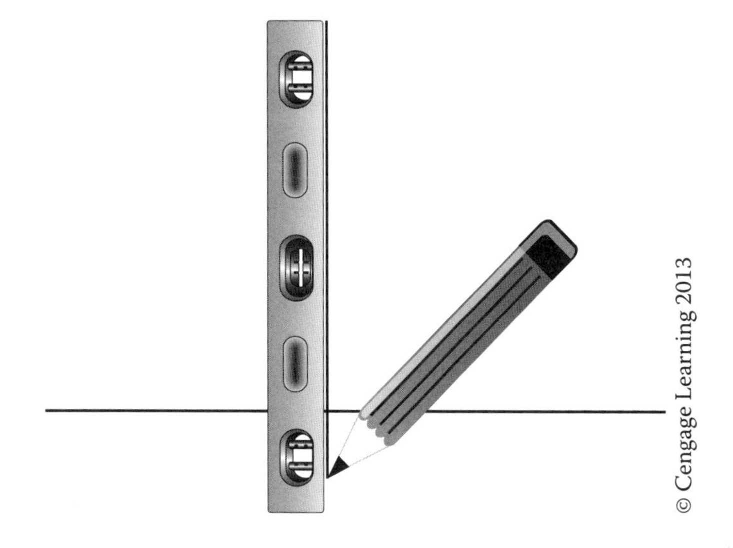

STEP 4: The brick's dimensions and the preferred head joint width determine the actual length of the sides. It is a common practice to have ⅜ inch-wide head joints. This length can be measured with a ruler to the nearest 1⁄16 inch increment or be *gauged*. The process of *gauging* allows accurate dimensioning without a ruler. After dry bonding the nominal size, 20 inch or 2½ inch brick for this illustration, the actual length is gauged or marked on the edge of a straight stick. A rectangle requires marking both the length and the width onto the gauge stick. Using a ruler to measure the lengths is optional.

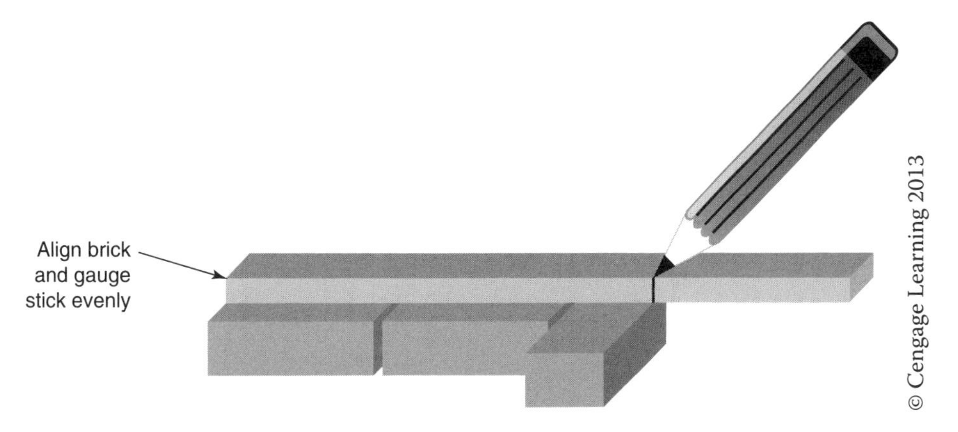

STEP 5: From the point of intersection, mark the length of the front side as gauged on the stick or measured with the ruler.

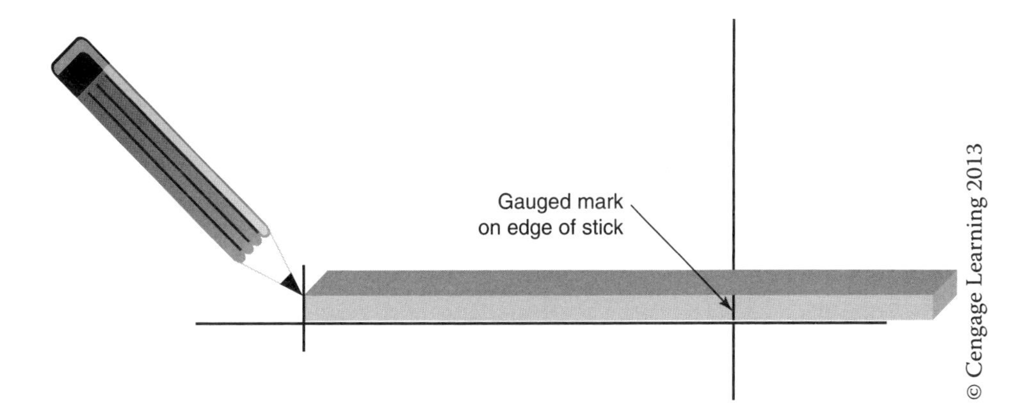

STEP 6: Mark the length of the right side using the gauged length or measured length.

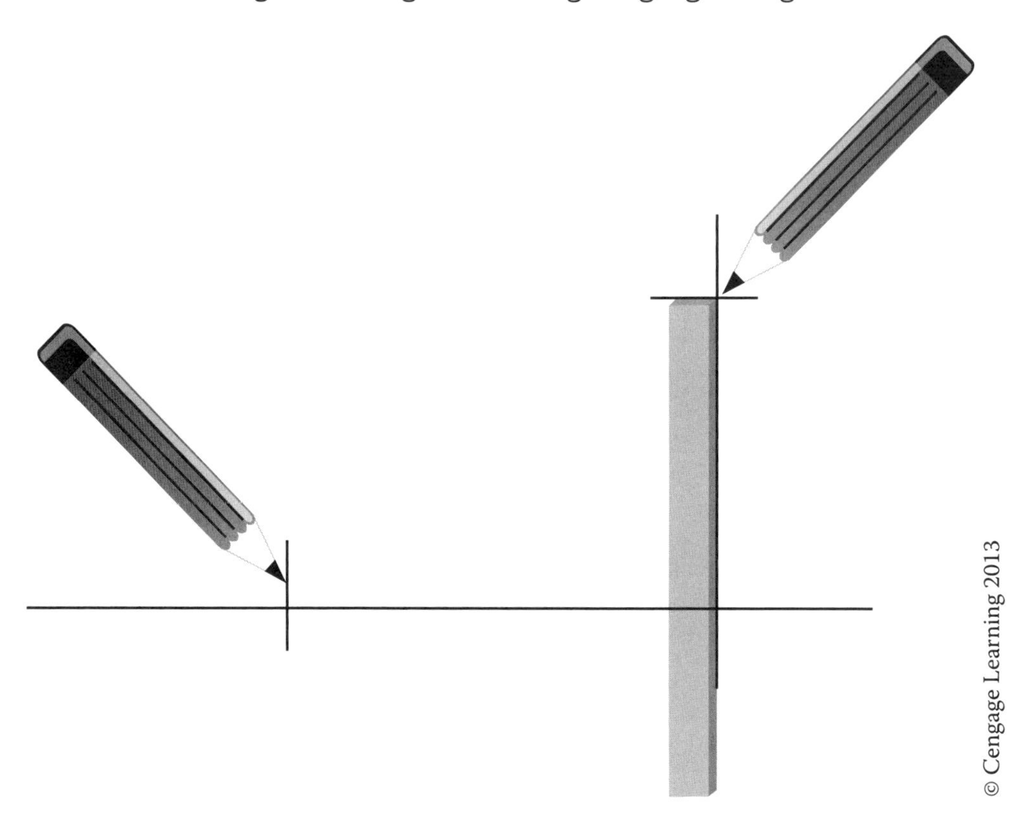

STEP 7: Having one blade of a framing square aligned parallel to the front line and the opposite blade aligned with the gauged length marking, construct a line at a right angle, perpendicular to the front side and parallel to the right side.

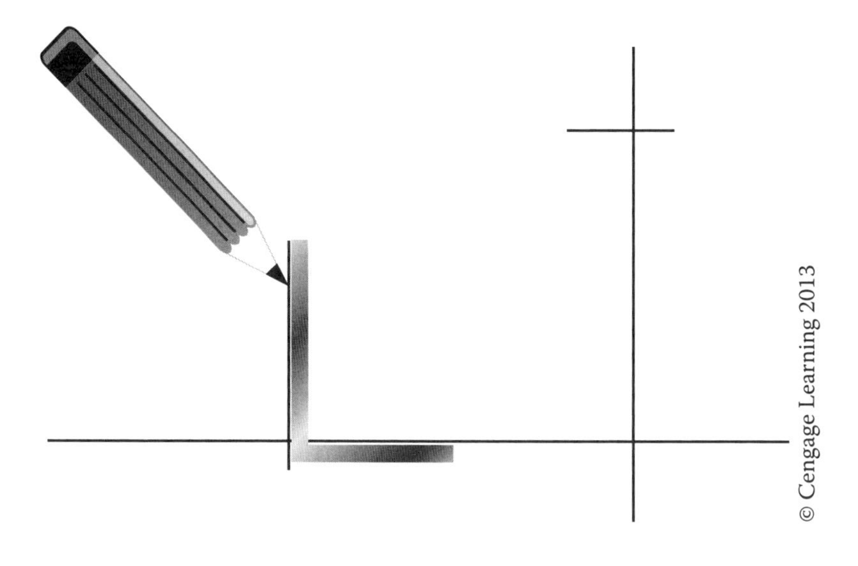

STEP 8: Using the edge of a 4 foot mason's level, lengthen this line. Continue the line further beyond its intersection with the first line.

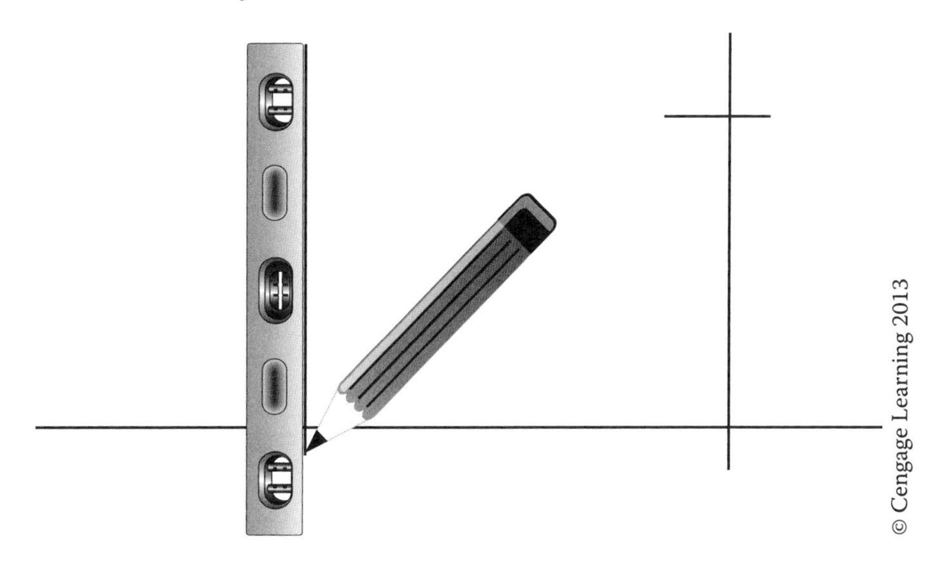

STEP 9: Mark the length of the left side using the gauge stick or the same measurement used for the front and right sides.

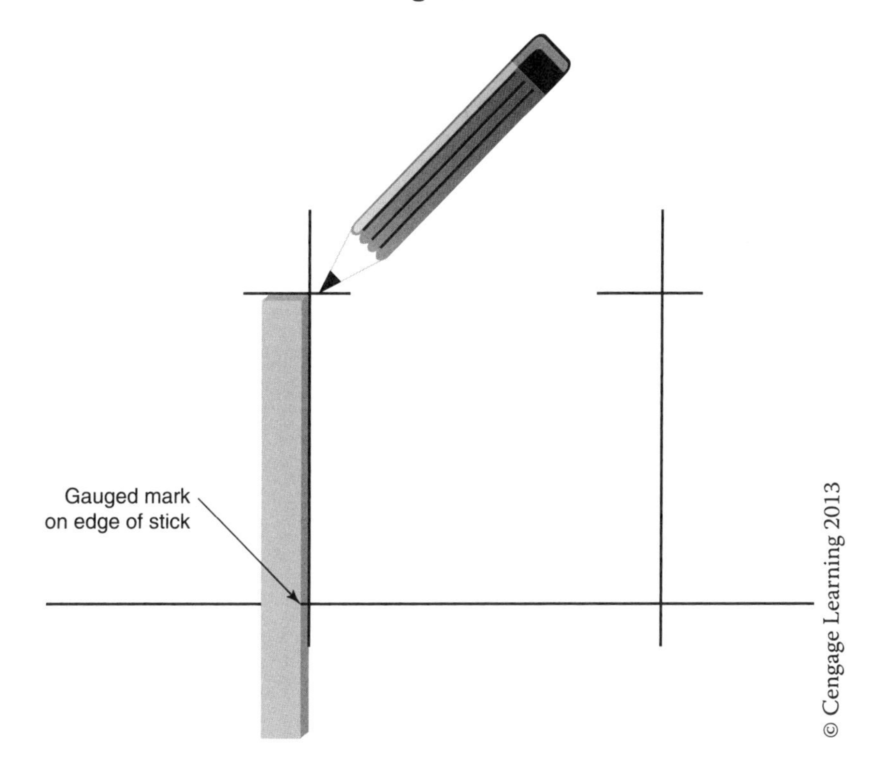

STEP 10: Using the straight edge of the 4 foot mason's level, draw a straight line at the point of the gauged markings between the opposite sides. Continue this line beyond its intersection with the previously drawn lines.

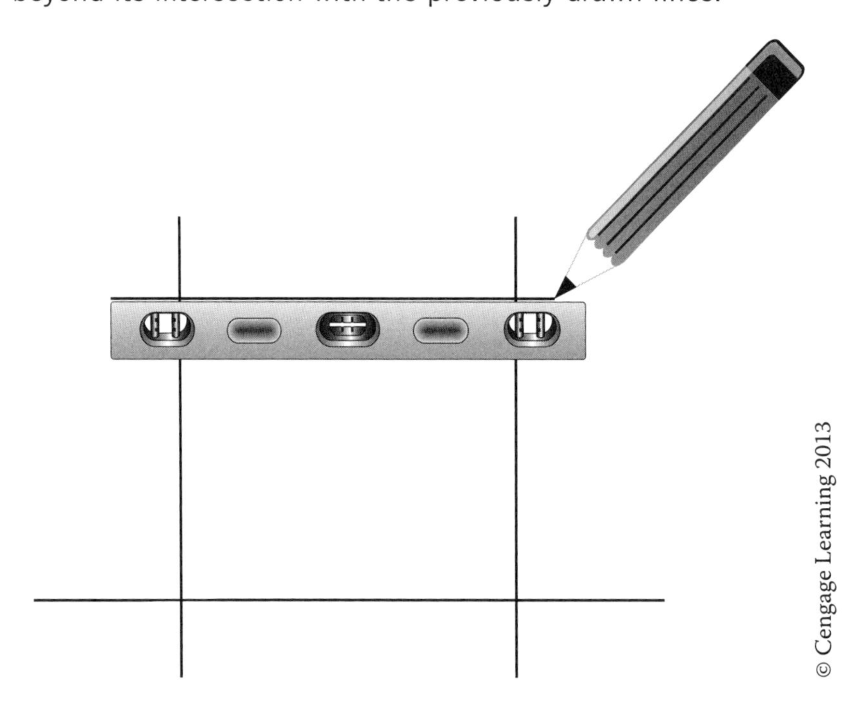

STEP 11: The square layout is now complete. The equal length of the diagonals (broken lines) confirms that the four sides are perpendicular. The diagonal measurements of rectangular columns and piers are also equivalent lengths.

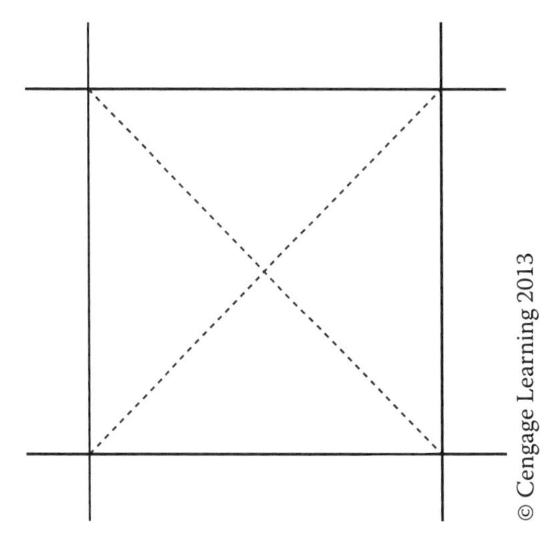

NOTE: It is a false assumption that a rectangular-shaped object is square if the front side length is equivalent to the back side length and the left side length is equivalent to the right side length. In the following illustration, the pairs of opposite sides are both parallel to one another and have equivalent lengths. Opposite angles are also equal. However, one can observe that adjacent sides are *not* perpendicular to one another and lengths of the diagonals are not equal. This shape is called a *parallelogram*.

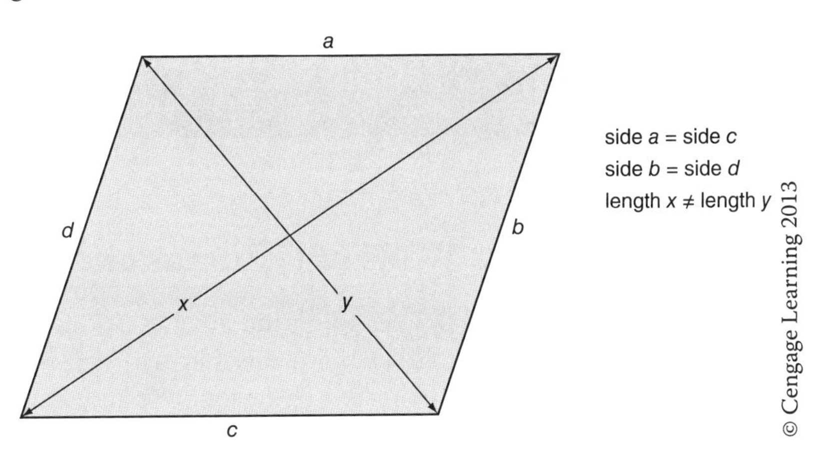

It is very important to compare the diagonal measurements of square- and rectangular-shaped projects to confirm square corners. The diagonal measurements should be equal.

Practical Problems

Solve each problem and round off all answers to the nearest hundredth (two places to the right of the decimal).

Using the procedures illustrated here, lay out a rectangle measuring 12″ × 18″ onto a flat surface. After confirming that all four sides are accurately dimensioned, measure the lengths of the two diagonals. If the rectangle is correctly laid out, each diagonal should have a length of 21.63 inches.

Lay out a rectangle measuring 4″ × 8″ onto a flat surface. Confirm measurements. The diagonals should measure 8.9 feet.

Lay out a square 4.5′ × 4.5′ once again on a flat surface. Measure the diagonals. They should measure 6.36 feet.

Lay out a rectangle 1′ × 10′. The diagonals measure 10.05 feet.

Lay out a rectangle measuring 44″ × 78″. Once again, after confirming that all four sides are accurate and all angles are 90°, measure the diagonals. The diagonals should measure 89.6 inches.

UNIT 32

Adjoining Walls of Columns and Piers

Basic Principles of Adjoining Walls of Columns and Piers

The following illustration is an 8 inch brick wall adjoining a 16″ × 16″ brick column at both ends. The columns should be aligned straight with one another with the wall between them. Parallel and perpendicular lines can be used to accurately lay out this and similar projects.

EXAMPLE 1: The following drawing illustrates how parallel and perpendicular lines can be used to lay out an 8 inch brick wall adjoining 16″ × 16″ brick columns at opposite ends. Here the adjoining wall between the columns has a length of 10 feet 8 inches. Using a chalk line reel, framing square, pencil, and rule, a mason can establish accurate layout lines on the supporting surface before the first brick is laid.

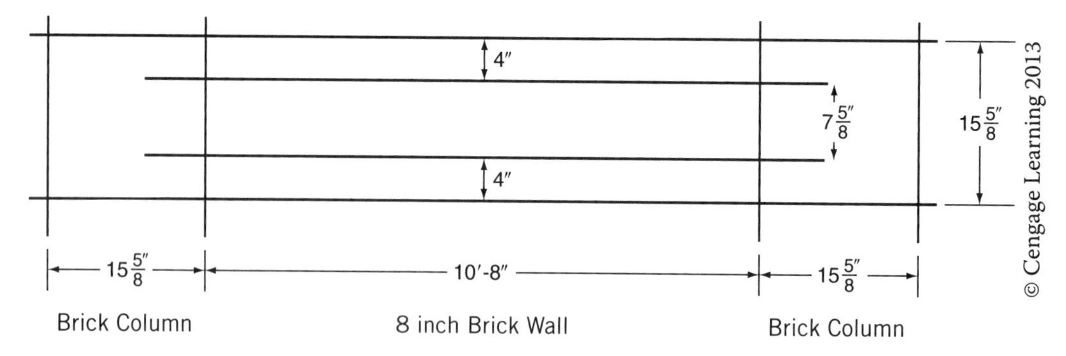

The shaded area in the following illustration identifies the "footprint" of the wall and end columns.

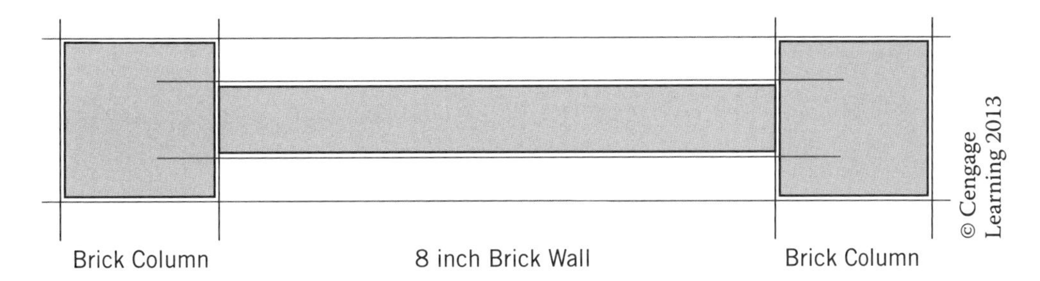

As seen in the following illustration, the addition of brick columns creates attractive brick garden walls. The columns not only add support to the wall but also give a stately presence to its form. Using parallel and perpendicular lines aids in the accurate layout of this and similar walls.

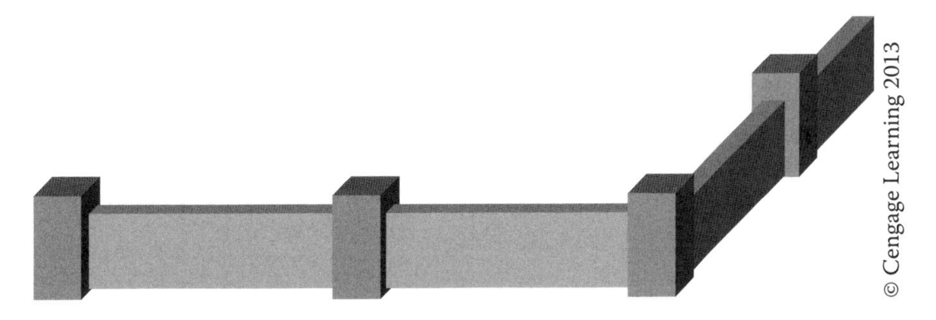

Did You Know?

Land surveyors can accurately locate all corners of a foundation by using a global positioning system and a defined reference point.

The use of parallel and perpendicular lines is also useful to lay out foundation walls having several corners. In the following illustration, the dashed lines represent the overall length and width of the foundation. By taking measurements from these lines, one can lay out all walls to have square corners. The dotted lines represent wall lines parallel to the dashed lines. They are established by taking equal measurements at opposite ends of the dashed lines, thereby having walls parallel with one another. This procedure enables one to mark or "pin" all corners of adjacent walls perpendicular to one another.

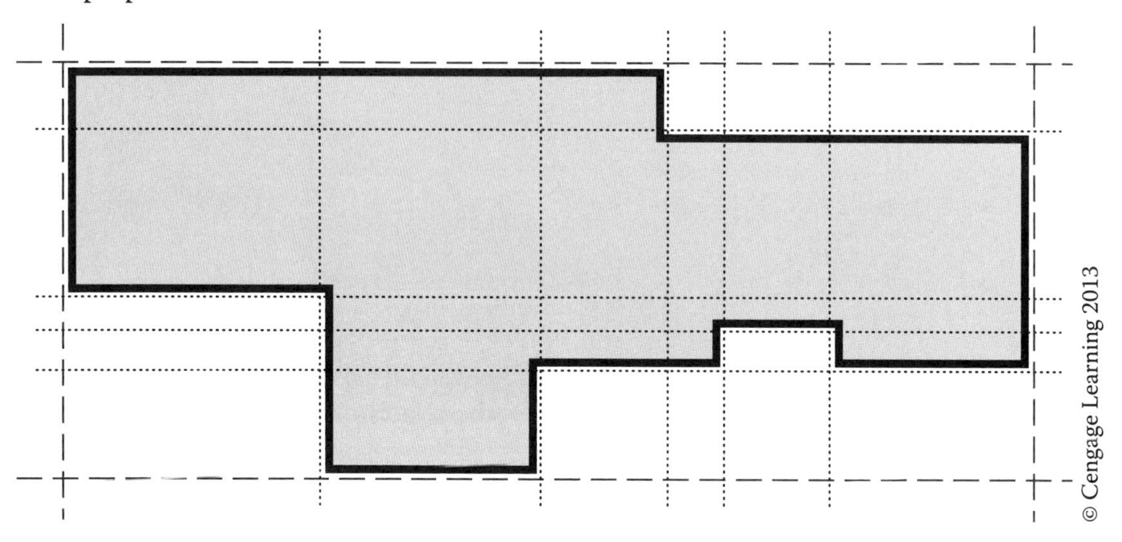

Practical Problems

1. The following illustration shows a driveway entry column and wall that is to be constructed using standard-size brick measuring 7⅝ inches long and 3⅝ inches wide. Mortar head joints are to be ⅜ inches wide. What is the correct spacing between the parallel and perpendicular layout lines? ______________

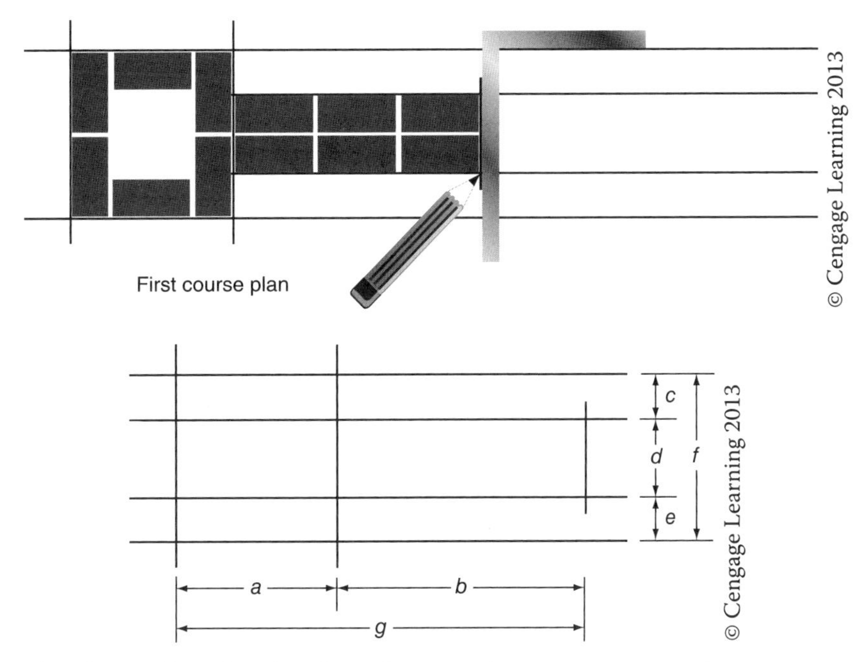

2. If a is three bricks and b is five bricks, what are the measurements?

a = ________″ e = ________″
b = ________″ f = ________″
c = ________″ g = ________″
d = ________″

3. If a is four bricks and b is eight bricks, what are the measurements?

a = ________″ e = ________″
b = ________″ f = ________″
c = ________″ g = ________″
d = ________″

UNIT 33

Pilasters and Chases

Basic Principles of Using Perpendicular and Parallel Lines to Align Pilasters and Chases with Adjoining Walls

Parallel and perpendicular lines are used to lay out *pilasters,* which are masonry columns projecting from masonry walls, and *chases,* which are recessed areas along the length of walls.

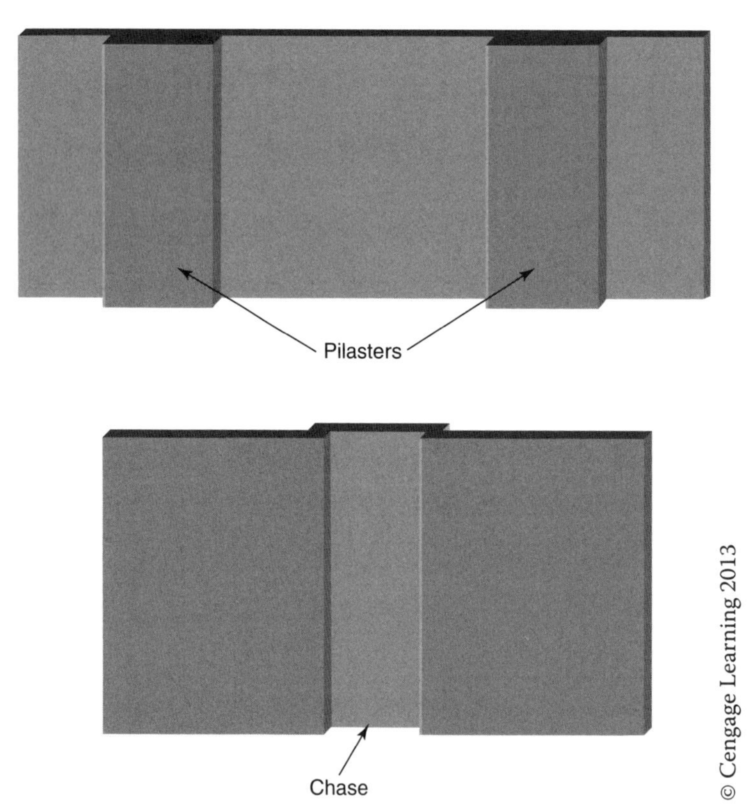

Properly aligned, the face of a wall and the face of a pilaster are parallel with each other. The two sides of the projecting pilaster are perpendicular to both its face and the face of the wall. The following illustration identifies the parallel and perpendicular lines necessary for proper alignment between a wall and pilaster.

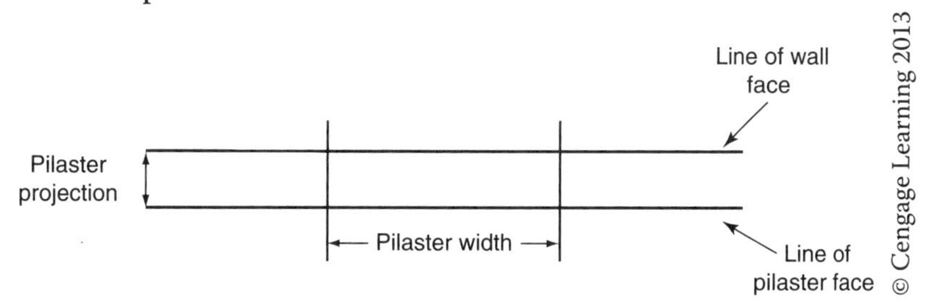

EXAMPLE 1: Look at the wall in the following instruction, which is an example of a pilaster when we face the wall from the bottom of the illustration, and a chase when we face the wall from the top of the illustration.

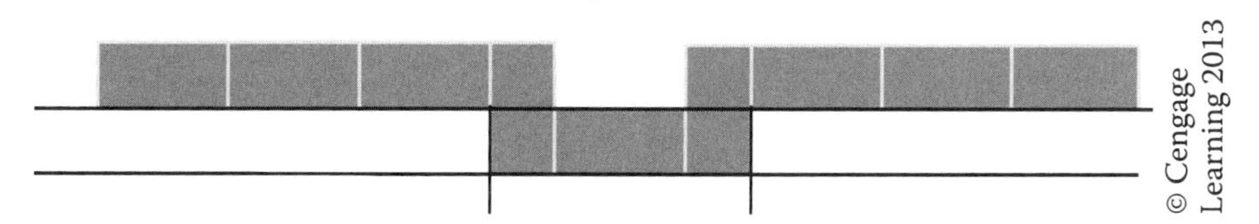

Practical Problems

1. Using standard-size brick measuring 7⅝ inches long by 3⅝ inches wide and ⅜ inch-wide mortar head joints, what is the correct spacing between the parallel and perpendicular layout lines?

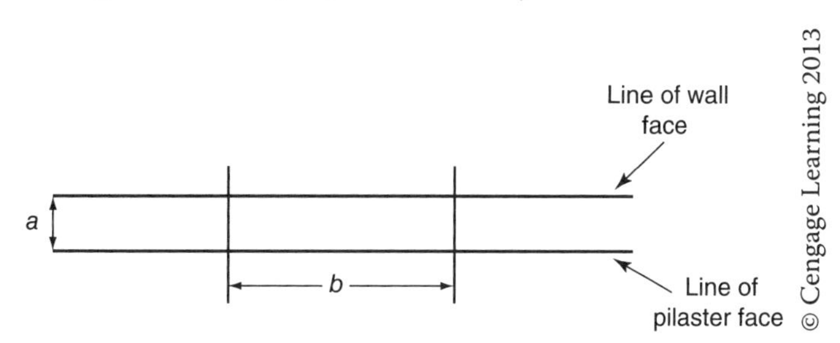

a = ________"

b = ________"

2. Using parallel and perpendicular lines enables masons to lay out chases also, a procedure similar to laying out pilasters. In the following illustration, how many inches deep is the chase, and what is its nominal width? ____________

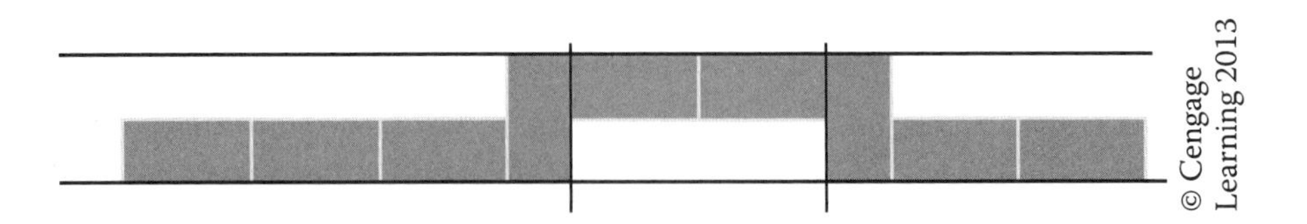

3. What is the overall length of the preceding wall? ____________

4. What is the overall width of the wall? ____________

5. If the depth of chase was 12 inches, how many more bricks would be needed per course in the preceding wall per course? ____________

6. In the preceding illustration, how many half bricks would be needed on the second course? ____________

UNIT 34

Square Corners for Foundation Walls

Basic Principles of Using the Pythagorean Theorem to Lay Out Square Corners for Foundation Walls

Although a carpenter's framing square is adequate for laying out and confirming square corners on smaller projects such as brick columns, piers, pilasters, and chases, the short lengths of a square's blades limit its use on larger projects. Mathematical calculations are necessary to lay out and confirm the square alignment of two adjacent foundation walls. The mathematical formula known as the Pythagorean Theorem can be used to lay out and confirm the square alignment of foundation walls (see Unit 23 for more on the Pythagorean Theorem). This theorem expresses a relation between the three sides of a right-angled triangle.

According to the Pythagorean Theorem: $a^2 + b^2 = c^2$

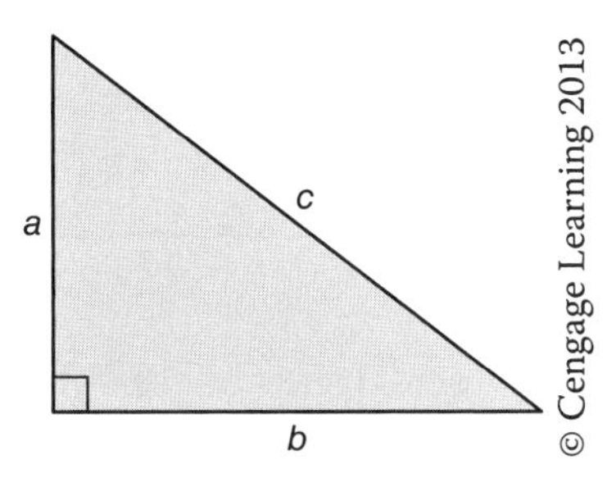

The right angle between adjacent sides can be confirmed for the following triangles by using the Pythagorean Theorem.

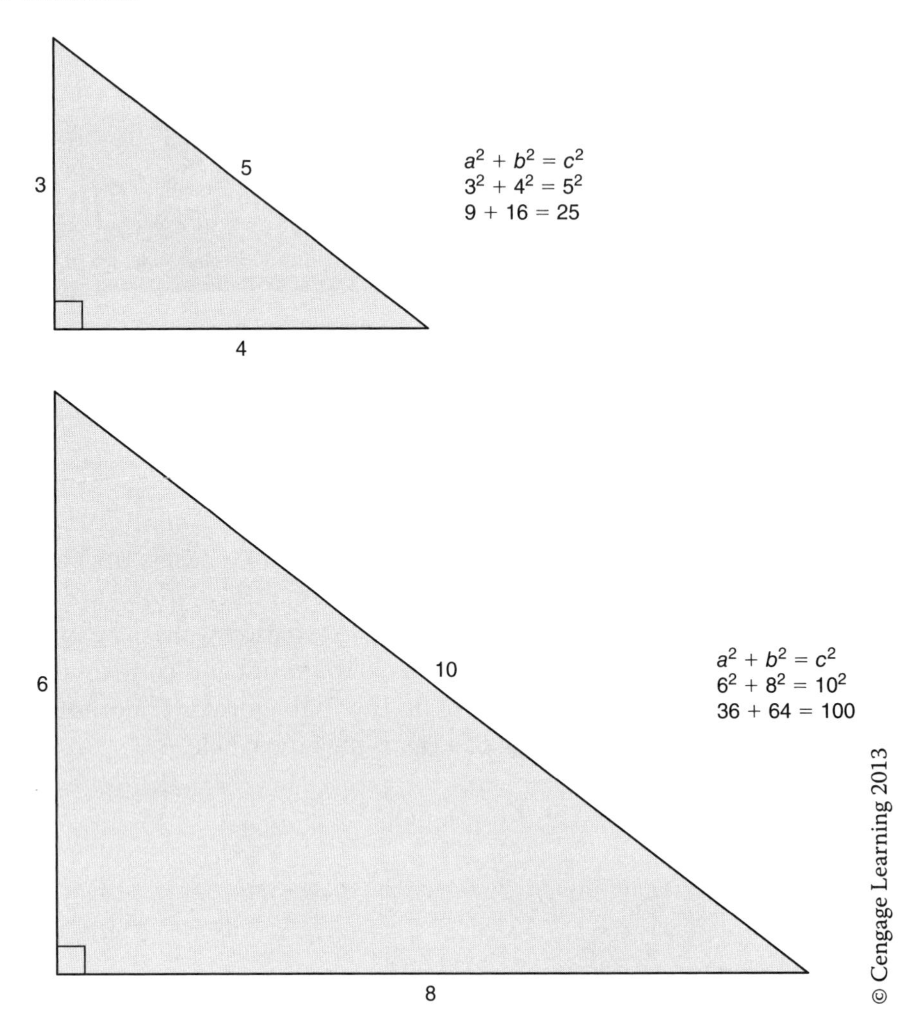

EXAMPLE 1: What should be the diagonal measurement reading (c) if two adjacent walls, one having a length of 28 feet and the other having a length of 48 feet, are aligned square?

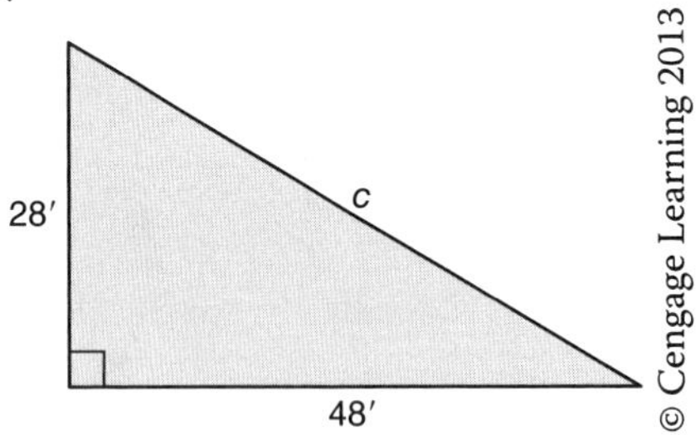

STEP 1: Set up the problem using the Pythagorean Theorem: $a^2 + b^2 = c^2$

$28^2 + 48^2 = c^2$

STEP 2: Calculate.

$c^2 = 784 + 2{,}304$
$c^2 = 3{,}088$

STEP 3: To find c, use the square root.

$c = \sqrt{3{,}088}$

$c = 55.57$

$c = 55'6\frac{12}{25}''$

ANSWER: The diagonal measure should be 55.57 feet or 55 feet $6\frac{12}{25}$ inches.

Practical Problems

Solve each problem and round off all answers to the nearest hundredth (two places to the right of the decimal). Give the equivalent measurement in feet, inches, and fractions of inches (sixteenths, eighths, quarters, or halves).

1. What should be the length of each diagonal measurement of a square column whose base dimensions are $15\frac{5}{8}'' \times 15\frac{5}{8}''$? ______________ inches (decimal equivalent) or ______________ inches (fractional equivalent)

2. What should be the length of each diagonal measurement of a $10' \times 12'$ patio if all four corners are square? ______________ feet (decimal equivalent) or ______________ feet ______________ inches

3. Two adjacent foundation walls have lengths of 26 feet 6 inches and 42 feet 8 inches. What should be the length of the diagonal measurement when these two walls are aligned perpendicular to one another? ______________ feet (decimal equivalent) or ______________ feet ______________ inches

4. If a 5′ × 7′ scaffold section is erected properly, braced on both sides of the end frames to provide a square setup, what should be the length of the diagonal measurement across the end frames? ______________feet (decimal equivalent) or ______________ feet ______________ inches

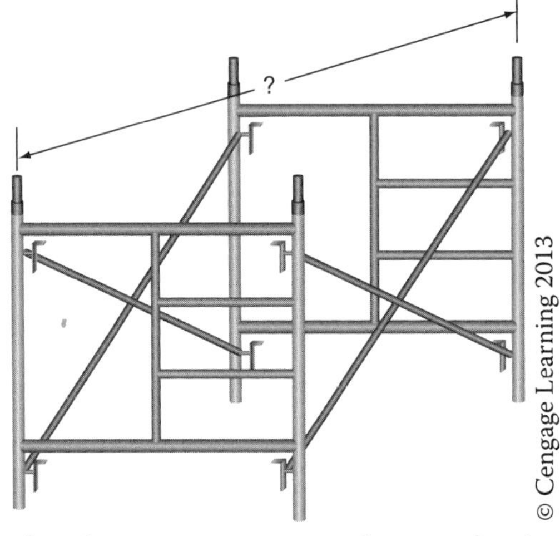

5. A flower box is 3.5′ × 18″ with square corners. What are the diagonals (in both fractional and decimal expression of inches)? ______________

6. A footing for a foundation measures 60′ × 40′ with square corners. What are the diagonals (in both fractional and decimal expression of inches)? ______________

SECTION 9

Materials Estimation

UNIT 35

Estimating Quantities of Face Brick

Basic Principles of Estimating Quantities of Face Brick

Estimating involves calculating the amount of materials needed for a job. In estimating, it is important that materials are not grossly underestimated or overestimated. Underestimating wastes time, energy, and money; overestimating is as time consuming and wasteful. Extra materials must be moved to another job or they might be damaged or eventually wasted.

Single-Wythe Brick Walls

The estimation of the quantity of bricks required to complete a single-wythe brick wall, one having a wall thickness of a single brick, is derived from multiplying the square feet of wall area by the number of bricks required per square foot. The number of bricks per square foot depends on the size of the brick. Including mortar joints and typical waste, there are approximately

- ❏ 6.75 standard-size bricks per square foot
- ❏ 5.8 engineered-size/oversize bricks per square foot
- ❏ 3 economy-size bricks per square foot

Wall openings such as doors and windows must be factored into the calculations or material estimates can be too high. For each wall, subtract the number of square feet represented by all openings from the square feet derived from the dimensions of the wall.

EXAMPLE 1: Determine the number of bricks needed to build a single-wythe brick wall 24 feet long and 10 feet high if the wall has one door opening measuring 3′ × 6′8″ and two window openings each measuring 3′ 4″× 4′.

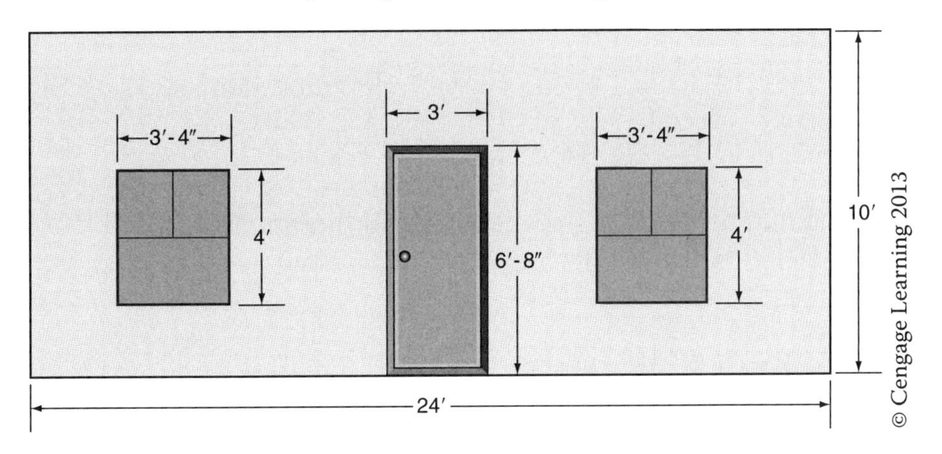

STEP 1: Find the total wall area using the formula area (A) = lw.

A = (24′)(10′)

A = 240 square feet

STEP 2: Find the total area of the wall openings.

area of door opening = (3′)(6′8″) = 19.98 square feet

area of each window opening = (3′4″)(4′) = 13.32 square feet

area of two window openings = (13.32)(2) = 26.64 square feet

total area of wall openings = 19.98 + 26.64 = 46.62 square feet

STEP 3: Subtract the combined area of the wall openings from the total wall area.

240 square feet − 46.62 square feet = 193.38 square feet (This is the wall area requiring bricks.)

STEP 4: Multiply the number of square feet requiring bricks by the number of bricks per square foot.

$(193.38)(6.75) = 1{,}305.3$ standard-size bricks

$(193.38)(5.8) = 1{,}121.6$ engineered-size/oversize bricks

$(193.38)(3) = 580.14$ economy-size bricks

ANSWER: The number of bricks needed to build the wall is 1,305.3 (standard); 1,121.6 (engineered); or 580.14 (economy).

Multiple-Wythe Brick Walls

For multiple-wythe brick walls, walls comprised of two or more bricks back-to-back, the estimated number of bricks is derived by first multiplying the square feet of wall area by the number of wythes and then multiplying this product by the number of bricks required per square foot.

EXAMPLE 2: How many engineered-size/oversize bricks are needed to build a double-wythe, 8 inch brick wall whose length is 36 feet and height is 32 inches?

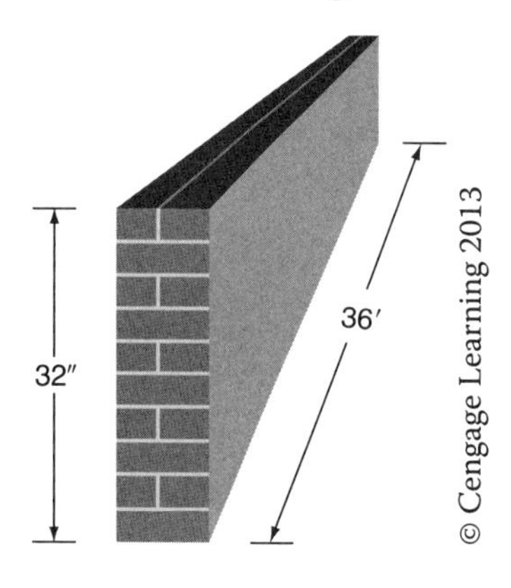

STEP 1: Find total wall area using the formula area $(A) = lw$

$A = (36')(2'8'')$

$A = 95.76$ square feet

 **STEP 2:** Multiply the number of square feet found in Step 1 by the number of wythes (in this case, 2) to calculate the combined square feet for all wythes.

$(95.76)(2) = 191.52$ square feet

STEP 3: Multiply the square feet of wall area (from Step 2) represented by the combined wall wythes by 5.8, which is the number of engineered-size/oversize bricks per square foot.

$(191.52)(5.8) = 1{,}110.82$ engineered-size/oversize bricks

ANSWER: This wall requires 1,110.82 engineered-size/oversize bricks.

Brick Rowlocks

Laid in a rowlock position, the back side of bricks is bedded in mortar. Brick rowlocks, or groupings of bricks laid in the rowlock position, appear as sills below windows and doors, as wall caps atop multiple-wythe brick walls, as masonry stair treads, and as fireplace hearths and mantle tops. More common rowlocks are the 8 inch, 12 inch, and 16 inch rowlocks. The number of bricks per linear foot of wall length depends on the size of the brick. Including mortar joints for a rowlock whose bed depth is equivalent to the length of one brick, there are approximately

- ❑ 4½ standard-size bricks per linear foot
- ❑ 3¾ engineered-size/oversize bricks per linear foot
- ❑ 3 economy-size bricks per linear foot

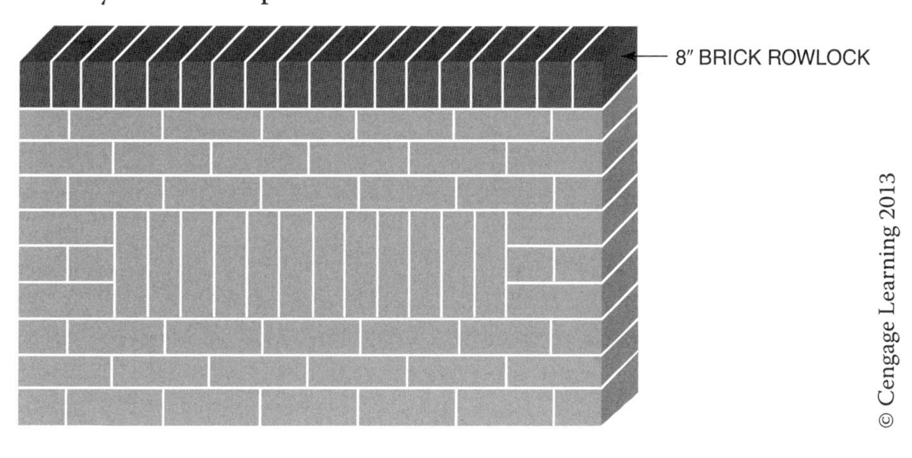

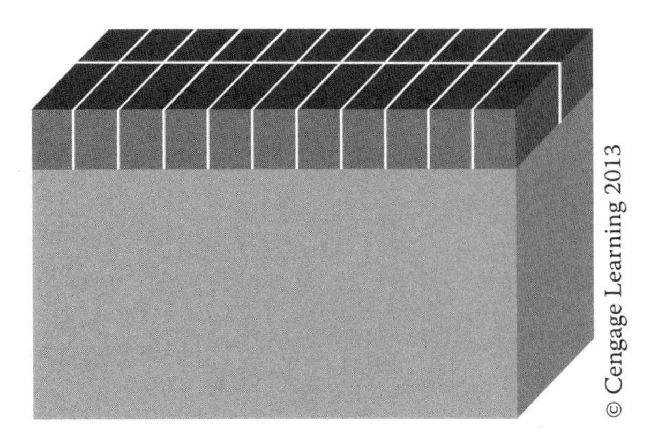

stacked pattern 12″ rowlock

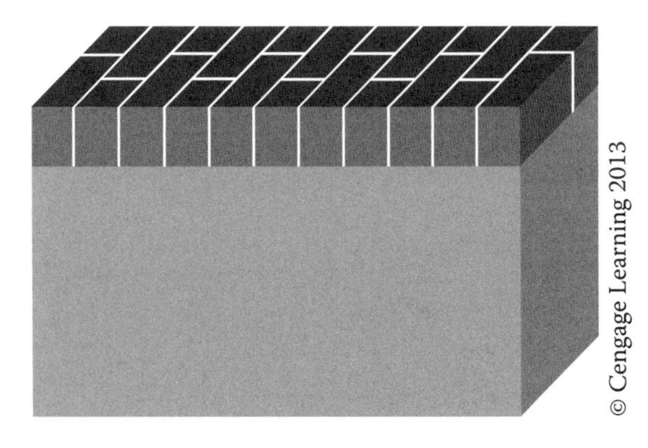

bonded pattern 12″ rowlock

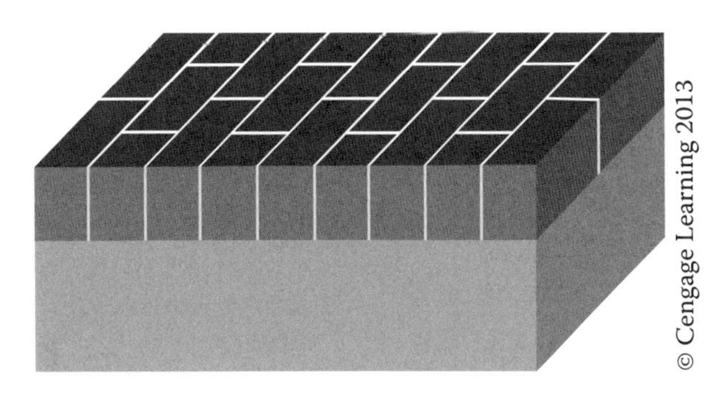

stacked pattern 16″ rowlock

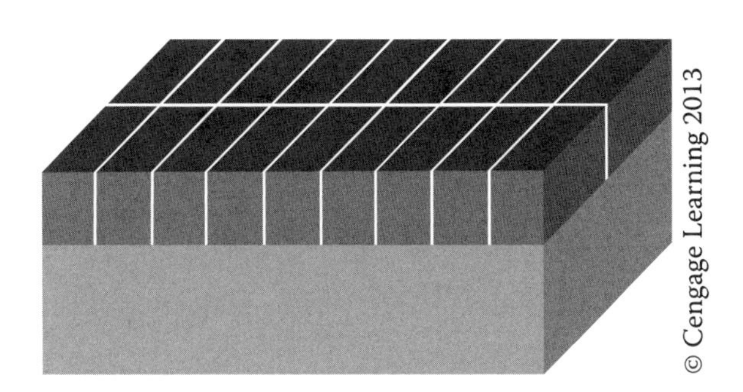

bonded pattern 16″ rowlock

EXAMPLE 3: How many standard-size bricks are needed for an 8 inch rowlock cap atop an 8 inch brick wall whose length is 4 feet?

STEP 1: Set up the problem.

(wall length)(number of bricks per linear foot) = number of bricks in the rowlock position

$(4)(4\frac{1}{2})$ = number of bricks

STEP 2: Multiply.

$(4)(4\frac{1}{2}) = 18$ bricks

ANSWER: 18 standard-size bricks are required for the rowlock cap.

Note: A 12 inch brick rowlock takes 1½ times as many bricks as an 8 inch rowlock and a 16 inch rowlock takes twice as many bricks as an 8 inch rowlock. Rowlocks topping the outer hearths of fireplaces are typically 2½ bricks wide, giving them the required extension of 20 inches, and they require 2½ times as many bricks as an 8 inch rowlock.

Estimating Bricks in Gable Areas

A gable end of a structure is the triangular area below the roofline. The formula for calculating the area of a triangle, ½bh, is used to calculate the area of a gable. For the equation, *b* equals the length of the gable and *h* equals the height of the triangle formed by the gable.

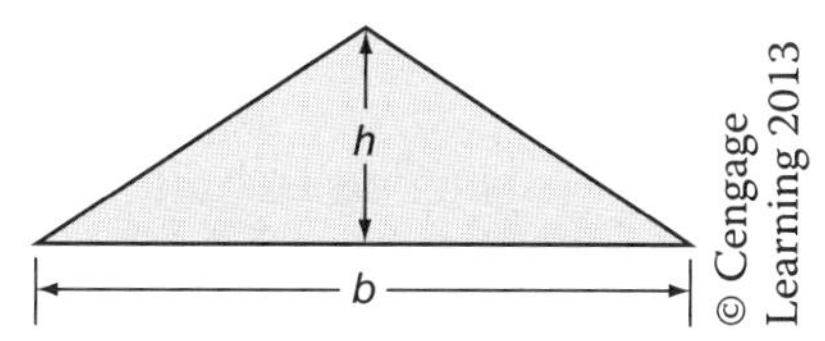

EXAMPLE 4: How many standard-size bricks are needed for this gable?

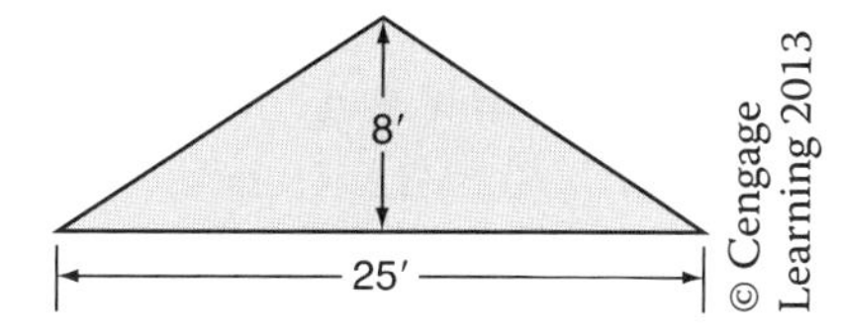

STEP 1: Set up the problem using the formula area (A) = ½bh

$A = \frac{1}{2}(25)(8)$

STEP 2: Multiply.

$A = \frac{1}{2}(200)$

$A = 100$ square feet

STEP 3: The number of standard-size bricks = (area)(6.75)

number of standard-size bricks = (100)(6.75)

number of standard-size bricks = 675

ANSWER: This gable will take an estimated 675 standard-size bricks.

The sum of the individual calculations for each wall of a given structure is an estimation of bricks required for the job.

Practical Problems

Solve each problem and round off all answers to the nearest hundredth (two places to the right of the decimal).

1. How many standard-size bricks are needed to build a single-wythe brick wall 52 feet long and 23 feet tall? ________________
2. How many engineered-size bricks are needed to build a double-wythe brick wall 24 feet long and 4 feet tall? ________________
3. How many economy-size bricks are needed to build a single-wythe brick wall 40 feet long and 16 feet tall? ________________
4. How many engineered-size bricks are needed to build a rowlock cap atop an 8 inch wall whose length is 16 feet 8 inches? ________________
5. How many standard-size bricks are needed to build a 12 inch-wide rowlock cap (a bed depth equivalent to the length of 1½ bricks) atop a retaining wall whose length is 37 feet? ________________
6. How many engineered-size bricks are needed to build a 20 inch-wide rowlock for a fireplace outer hearth whose length is 7 feet? ________________
7. How many standard-size bricks are needed to build a rowlock cap for a 16″ × 16″ brick pier? ________________

Refer to the following illustration to answer problems 8–10.

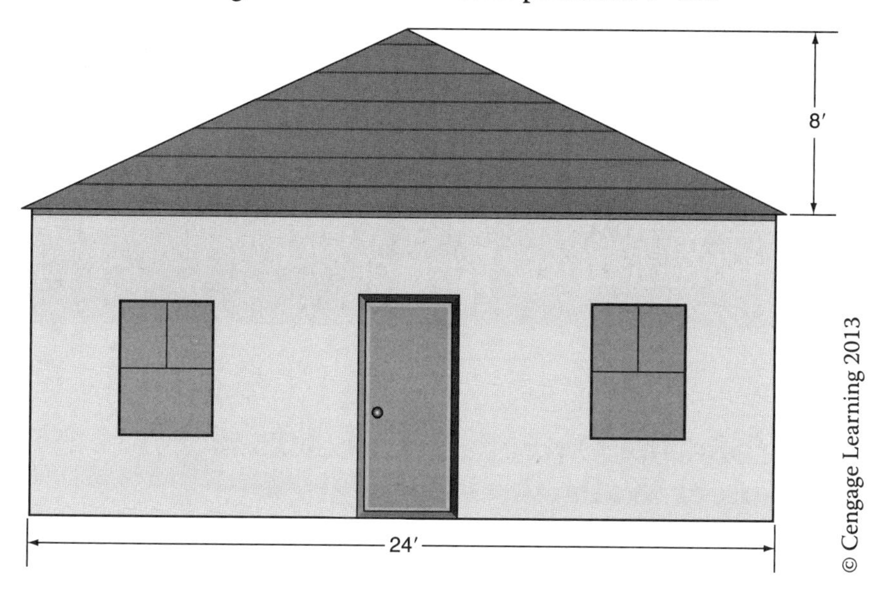

8. How many square feet comprise the gable area? __________
9. How many standard-size bricks are needed for the gable? __________
10. How many engineered-size bricks are needed for the gable? __________

UNIT 36

Estimating Quantities of Steel Ties for Anchored Brick Veneer

Basic Principles of Estimating Corrugated Steel Ties for Anchored Brick Veneer Walls

Corrosion-resistant, zinc-coated, corrugated steel ties anchor 4 inch brick veneer walls to wood framing. They are typically spaced 16 or 24 inches apart horizontally, the same as the spacing of wall studs, and secured with corrosion-resistant nails or screws to the wall studs. It is common practice to install wall ties every six courses of standard-size brick or every five courses of oversize bricks, a vertical spacing of approximately 16 inches. A 16″ × 16″ spacing results in a spacing of one wall tie for every 1.78 square feet and a 24″ × 16″ spacing is equivalent to a spacing of one wall tie every 2.67 square feet, the maximum allowable spacing. An equal number of corrosion-resistant nails or screws are needed to secure each wall tie with a single fastener penetrating the wall studs a minimum of 1½ inches.

A 16″ × 16″ spacing of wall ties is proportionate to one wall tie per 12.4 standard-size bricks or one wall tie per 9.7 engineered-size bricks. A 24″ × 16″ spacing of wall ties is proportionate to one wall tie per 18.7 standard-size bricks or one wall tie per 14.7 engineered-size bricks.

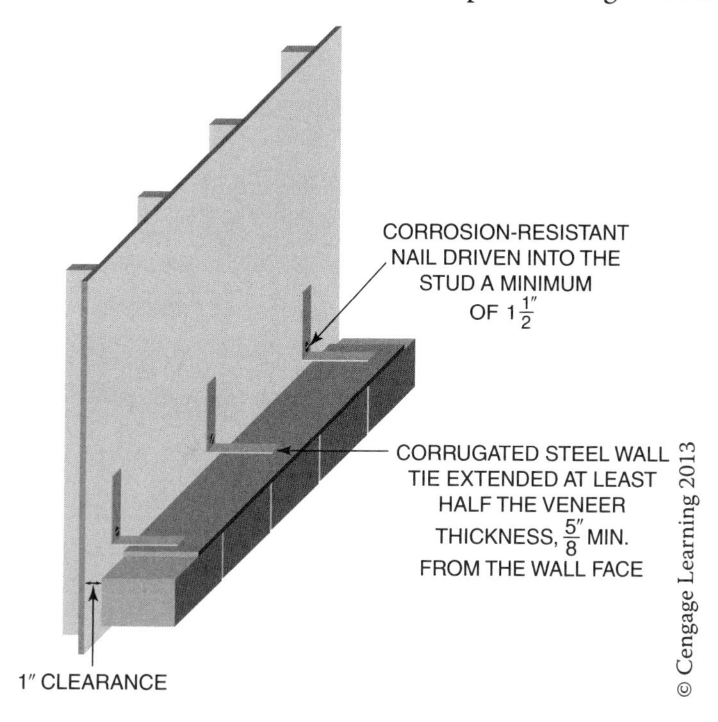

EXAMPLE 1: How many corrugated wall ties and fasteners are needed to anchor 350 square feet of brick veneer to wood stud framing using a 16″ × 16″ spacing for ties?

STEP 1: Divide the square feet of the wall area by 1.78, which is the square feet of wall area supported by a single wall tie for a 16″ × 16″ spacing.

$350 \div 1.78 = 196.6$, which can be rounded up to 197

ANSWER: Approximately 197 wall ties and 197 corrosion-resistant fasteners are needed.

Alternate method for solving this example:

STEP 1: Multiply the wall area by 6.75, which is equal to the number of standard-size bricks per square foot.

(350)(6.75) = 2,362.5 standard-size bricks

STEP 2: Divide the total number of bricks found in Step 1 by 12.4.

2,362.5 ÷ 12.4 = 190.52 rounded up to 191

ANSWER: Approximately 191 wall ties and 191 corrosion-resistant fasteners are needed.

Practical Problems

Solve each problem round off all answers to the nearest hundredth (two places to the right of the decimal).

1. How many wall ties are needed for a brick wall 44 feet long and 9 feet tall if the spacing of ties is 16″ × 16″? __________
2. How many wall ties are needed for a job requiring 21,500 engineered-size bricks if wall ties are spaced 16 inches apart both horizontally and vertically? __________
3. How many wall ties are needed for a wall measuring 64′ × 12′8″ if the maximum allowable spacing of one wall tie per 2.67 square feet is observed? __________
4. How many wall ties are needed for a wall measuring 36 feet long and 8 feet tall if the spacing of ties is 24″ × 16″? __________
5. How many wall ties and fasteners are needed to anchor 265 square feet of brick veneer to wood stud framing using a 16″ × 16″ spacing for ties? __________
6. Using a spacing of one wall tie for every 1.78 square feet, how many wall ties are needed for a wall measuring 48′ × 10′6″? __________

UNIT 37

Estimating Quantities of Blocks or CMUs for Wall Construction

Basic Principles of Estimating Blocks or CMUs for Wall Construction

Estimating concrete masonry units, often referred to as CMUs or blocks, can be accomplished using the area method. Regardless of block size, the face side of full-size blocks has the same length and height, that being 15⅝ inches long and 7⅝ inches high. Block *size* defines the block's nominal width. The quantity of blocks required to build a given wall is the same regardless of the block size. A block's face, including a ⅜ inch-wide mortar bed joint and head joint, measures 16″ × 8″. This is equivalent to an area of 128 square inches.

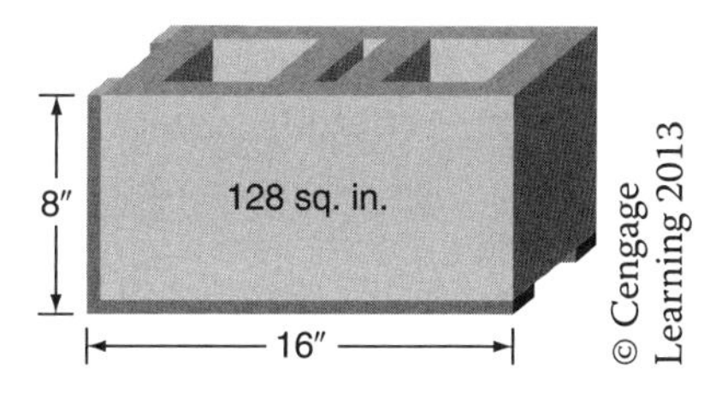

Since there are 144 square inches per square foot, one block is equivalent to 128/144th of 1 square foot, or 0.88 square foot. Put another way, it takes 1.125 blocks to complete 1 square foot of wall area.

The product resulting from multiplying the square feet of the wall area by 1.125 equals the number of blocks required to build a given wall.

EXAMPLE 1: How many blocks are needed to construct a wall having no wall openings that measures 25 feet long and 8 feet high?

STEP 1: Determine the total wall area using the formula area (A) = lw

$A = (25')(8') = 200$ square feet

STEP 2: Multiply the area by 1.125 to determine the number of blocks.

$(200)(1.125) = 225$ blocks

ANSWER: Approximately 225 blocks are needed for this wall.

As when estimating bricks, subtract the combined area of wall openings from the overall wall area *before* multiplying by 1.125. Otherwise, estimations will be greater than necessary. The sum of the individual calculations for each wall of a given structure can be used as an estimation of blocks required for a specific job. However, the method used in the following sample problem eliminates calculating materials for each individual wall. Rather, it relies upon finding the perimeter of structures such as foundation walls. Note that the actual block count may be slightly less than the estimation because the block width at corners is not taken into account.

EXAMPLE 2: Determine how many blocks are needed to build a foundation having dimensions as illustrated in the following working drawing. The height of all foundation walls is 9 feet 4 inches. Wall openings include one $3' \times 6'8''$ doorway and three window openings each measuring $3' \times 4'$.

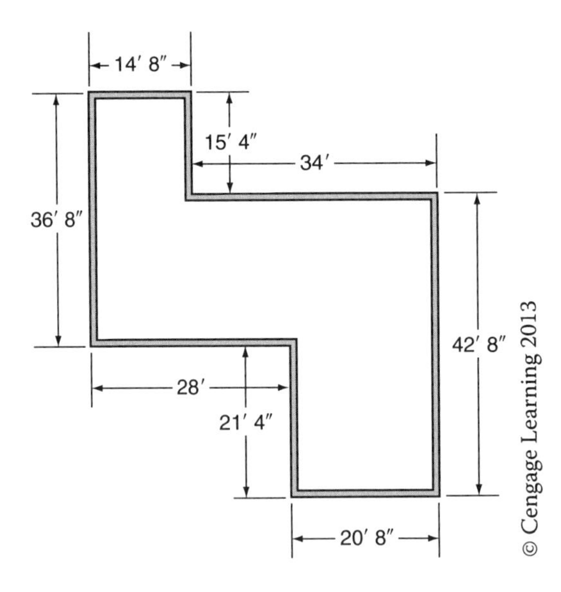

STEP 1: Find the perimeter length of the foundation. The perimeter (P) is equal to the sum of the length of every wall.

P = combined length of all sides

$P = 14'8'' + 15'4'' + 34' + 42'8'' + 20'8'' + 21'4'' + 28' + 36'8''$

$P = 213'4''$

STEP 2: Find the total wall area. The total wall area (A) is equal to the perimeter length (P) multiplied by the wall height.

$A = lw$

$A = (213'4'')(9'4'') = 1{,}991.11$ square feet

STEP 3: Find the area (A) of all wall openings. The area of each wall opening is equal to its length multiplied by its height.

area (A) of doorway opening = lw

$A = (3')(6'8'')$

$A = 19.98$ square feet

area (A) of 3 identical size window openings = (3)lw

$A = (3)(3)(4)$

$A = 36$ square feet

the combined area (A) of 1 door opening and 3 window openings =

19.98 square feet + 36 square feet = 55.98 square feet

STEP 4: Subtract the area of the wall openings from the total wall area (A) from Step 2. The remaining area is equal to the wall surface area requiring blocks.

$1{,}991.11 - 55.98 = 1{,}935.13$ square feet

STEP 5: Calculate the estimated quantity of blocks needed by multiplying the square feet of remaining wall area by 1.125.

$(1{,}935.13)(1.125) = 2{,}177.02$ blocks

ANSWER: Approximately 2,177.02 blocks

Practical Problems

Solve each problem and round off all answers to the nearest hundredth (two places to the right of the decimal).

1. This foundation measures 44 feet in width and 66 feet in length. The single-wythe foundation walls are 8 feet high. How many CMUs are needed to construct these walls? _______________

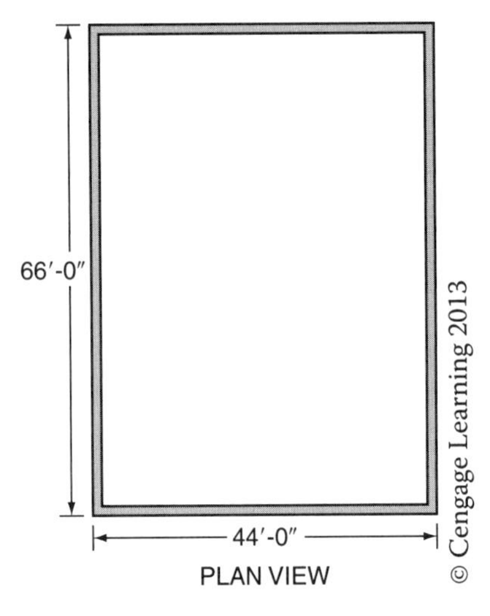

2. Determine how many CMUs need to be ordered for a concrete building 32 feet wide by 52 feet long, constructed with 8-foot-high walls. Wall openings include two doorways each measuring 32″ × 6′8″ and two windows each measuring 4′ × 2′. _______________

3. If a storage building measures 100 feet in length and 60 feet in width, with 20 foot-high walls, how many CMUs would be required if there are two 8′ × 4′ windows and two 10′ × 12′ doors? ______________

4. If the height of the walls of a building is 10 feet, and the building is 48′ × 36′, with 148 square feet of wall openings, how many CMUs are needed? ______________

5. A garage measuring 24′ × 24′ has walls 8 feet high. There are four windows, each measuring 3′ × 3′, and two sets of double garage doors each measuring 7′6″ × 7′6″. How many CMUs are needed to build the garage? ______________

6. A building measuring 44′ × 20′ with 8-foot-high walls has 36 square feet of wall openings.

   a. What is the net wall area, the total wall area minus the area of all wall openings? ______________

   b. How many CMUs are needed for the building? ______________

UNIT 38

Estimating Quantities of Bagged Masonry Cement for Brick and Block Wall Construction

Basic Principles of Quantities of Bagged Masonry Cement for Brick and Block Wall Construction

The ingredients of masonry cements, mortar cements, and Portland Cement-Lime (PCL) mixes are accurately proportioned, blended, and bagged by the manufacturers. Typically packaged in bags holding a volume of 1 cubic foot for easy handling at the jobsite, one such bag is sufficient for making mortar to lay in the running bond pattern approximately 125 standard-size bricks, 100 oversize/engineered-size bricks, 80 economy/utility-size bricks, or 40 blocks.

The quotient derived by dividing the total number of masonry units needed for a specific job by the estimated number of units that can be laid with one bag of masonry cement is an estimate of the number of 1-cubic-foot–size bags of masonry cement required for the mortar joints of the corresponding masonry unit. Actual mortar requirements for bricks may increase by 10% or more depending on the mortar joint widths, the bond pattern, the size of brick cores, and mortar management. Factors determining actual mortar requirements for block work include mortar joint widths, solid bedding versus face shell spreading below the course, and mortar management.

EXAMPLE 1: How many 1-cubic-foot–size bags of masonry cement are needed to lay 1,000 standard-size bricks in the running bond pattern?

STEP 1: Divide.

$1,000 \div 125 = 8$

ANSWER: Approximately eight bags of masonry cement are needed.

Practical Problems

Solve each problem and round off all answers to the nearest hundredth (two places to the right of the decimal).

1. How many 1-cubic-foot–size bags of masonry cement are needed to lay 160 blocks? ________________

2. Determine the number of 1-cubic-foot–size bags of masonry cement required to lay 800 oversized/engineered-size bricks in the running bond pattern. ________________

3. How many 1-cubic-foot–size bags of masonry cement does it take to lay 37,300 standard-size bricks in the running bond pattern? ________________

4. How many 1-cubic-foot–size bags of masonry cement are needed to lay 1,500 economy-/utility-size bricks in the running bond pattern? ________________

5. Determine how many 1-cubic-foot–size bags of masonry cement are needed to lay 3,750 blocks. ________________

UNIT 39

Estimating Cubic Yards of Masonry Sand for Making Brick and Block Mortar

Basic Principles of Estimating Cubic Yards of Masonry Sand for Making Brick and Block Mortar

Masonry sand is mixed with 1-cubic-foot–size bags of masonry cement, mortar cement, or PCL mixes whose ingredients are blended and accurately proportioned by manufacturers to make mortar for laying bricks and blocks. The sand is measured by volume to provide constant results from batch to batch, ensuring uniform color and strength. The ratio of masonry cement, mortar cement, or PCL mixtures to sand is typically one part cement to three parts sand. This ratio can be the same for mixing mortar for laying brick or block. One cubic yard of damp sand is needed for every eight bags of cement when mixing the two in a 1:3 ratio (one part cement and three parts sand) to make mortar. Dividing the required number of 1-cubic-foot–size bags of cement by 8 provides an estimate of the number of cubic yards of masonry sand needed.

EXAMPLE 1: How many cubic yards of sand are needed to mix 72 bags of masonry cement to make mortar?

STEP 1: Set up the problem using the formula cubic yards of sand = number of bags of cement ÷ 8.

cubic yards = 72 ÷ 8

STEP 2: Divide.

cubic yards = 72 ÷ 8 = 9

ANSWER: Approximately 9 cubic yards of sand are needed.

Practical Problems

Solve each problem and round off all answers to the nearest hundredth (two places to the right of the decimal).

1. How many cubic yards of sand are needed to mix with 360 bags of masonry cement to make mortar? __________
2. If a job requires 48 1-cubic-foot–size bags of masonry cement, how many cubic yards of sand will be needed to make mortar? __________
3. If a job requires 12 cubic yards of sand to make mortar, how many 1-cubic-foot–size bags of masonry cement will be used? __________
4. How many cubic yards of sand are needed for a job requiring the use of 268 bags of masonry cement of 1 cubic foot size? __________
5. How many 1-cubic-foot–size bags of masonry cement can be mixed using 38 cubic yards of sand? __________

UNIT 40

Estimating Cubic Yards of Grout for Reinforcing Block Walls

Basic Principles of Estimating Cubic Yards of Grout for Reinforcing Block Walls

Composite and cavity walls can be reinforced with steel bar reinforcement and grout. Grout is a mixture of cement and aggregates to which sufficient water is added to produce a pouring consistency. Block walls are strengthened by adding grout to their hollow block cells. The approximate calculated volumes of grout required per 100 square feet of block wall surface area are:

- ❑ 1.85 cubic yards for 12 inch block
- ❑ 1.27 cubic yards for 10 inch block
- ❑ 1.10 cubic yards for 8 inch block
- ❑ 0.77 cubic yards for 6 inch block

EXAMPLE 1: How many cubic yards of grout are needed to fill an 8 inch block wall having a length of 16 feet and height of 10 feet?

STEP 1: Find the area of the wall using the formula area (A) = lw.

$A = (16')(10') = 160$ square feet

STEP 2: Divide the square feet of wall area by 100 since grout quantities in this chart represent quantities per 100 square feet of wall area.

$160 \div 100 = 1.6$

STEP 3: Multiply the factor derived in Step 2 by the quantity of grout per 100 square feet as given in the preceding list.

$(1.6)(1.10) = 1.76$ cubic yards grout

ANSWER: Approximately 1.76 cubic yards of grout is needed.

Practical Problems

Solve each problem and round off all answers to the nearest hundredth (two places to the right of the decimal).

1. How many cubic yards of grout are needed to fill a 6 inch block wall with a length of 26 feet and height of 10 feet? ________________
2. If a wall measures $15' \times 36'$, and is made of 8 inch blocks, how many cubic yards of grout are needed to fill all block cells? ________________
3. Determine how many cubic yards of grout are needed to fill two 12 inch block walls each measuring 18 feet in length and 12 feet in height. ________________
4. How many cubic yards of grout are needed to fill a 10 inch block wall having a length of 24 feet and height of 8 feet? ________________
5. How many cubic yards of grout are needed to fill a 12 inch block wall measuring $15' \times 10'6''$? ________________

UNIT 41

Estimating Quantities of Anchor Bolts for Foundation Walls

Basic Principles of Estimating Quantities of Anchor Bolts or Anchor Straps for CMU Foundation Walls

Code-compliant job specifications dictate the spacing of threaded foundation anchor bolts or sheet-metal straps used for securing sill plates, (2-inch-thick framing lumber used to support perimeter band joists and floor joists) to the tops of foundation walls. Depending on governing regulations, a maximum spacing of 6 feet along the length of foundation walls and not more than 12 inches from each corner may be acceptable. To determine how many anchor bolts are needed for a single wall, divide its length by the required spacing, rounding up quotients to the next whole number, and adding 1 to the rounded up quotient.

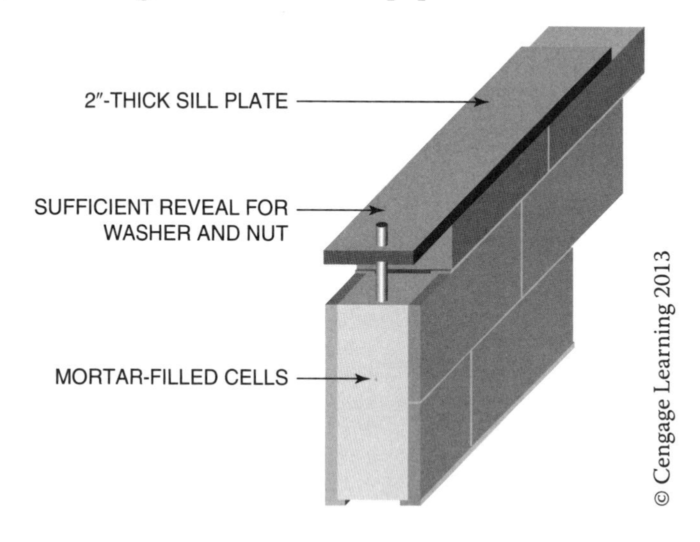

EXAMPLE 1: How many anchor bolts are needed for a 28 foot-long block wall if the spacing between anchor bolts is 6 feet?

STEP 1: Divide the length of the wall by 6, the maximum spacing for anchor bolts.

$28 \div 6 = 4.66$ rounded up to 5

STEP 2: To the rounded up quotient in Step 1, add 1 for the additional bolt needed at the beginning end of the wall.

$5 + 1 = 6$

ANSWER: Approximately six anchor bolts are needed.

The total number of foundation bolts needed for an entire foundation is equivalent to the sum of the anchor bolts needed for each wall of the foundation.

Practical Problems

Solve each problem and round off all answers to the nearest hundredth (two places to the right of the decimal).

1. How many anchor bolts are needed for a block wall 32 feet long if the spacing between anchor bolts is 4 feet? __________
2. For a 42-foot-long block wall, how many anchor bolts are needed if the spacing between anchor bolts is 6 feet? __________
3. If the spacing between anchor bolts is 3 feet, how many anchor bolts are need for a block wall 18 feet long? __________
4. How many anchor bolts are needed for a block wall 56 feet long if the spacing between anchor bolts is 3 feet 6 inches? __________
5. How many anchor bolts are needed for a block wall 22 feet long if the spacing between anchor bolts is 4 feet? __________

UNIT 42

Estimating Cubic Yards of Concrete for Perimeter Footings

Basic Principles of Estimating Cubic Yards of Concrete for Foundation Footings

Cubic yard is the unit of measure used to calculate the quantity of concrete required for placing footings, which are the structural support for most masonry walls. Cubic yards of concrete, referred to in the trade as simply *yards*, are determined using the following formula.

cubic yards of concrete (cu yd) = (l′w′h′) ÷ 27

Length (l) equals the length of a wall or the sum of the lengths of multiple walls supported by the footing. *Width (w)* equals the width of the footing.

This mathematical formula requires both length (l) and width (w) to be expressed as *feet*. For example, to build a 12 inch block wall that is 24 feet 6 inches long, length (l) is 24.5′. Width (w) for residential construction is usually the greater of (1) a minimum of twice the width of the block width, or (2) 2 feet.

The constant 0.66 represents a thickness of 8 inches, equivalent to 8⁄12 feet, or 0.66 feet, a thickness often specified for footings supporting foundation walls in residential construction. Multiplying length (l) × width (w) × 0.66 calculates the volume of concrete as cubic feet. Since 27 cubic feet is equivalent to 1 cubic yard, dividing the number of cubic feet by *27* converts cubic feet calculations to cubic yards, which is the unit of purchase for truck-delivered, ready-mixed concrete.

EXAMPLE 1: A garage 36 feet long and 24 feet wide is to be constructed using 8 inch blocks. The concrete footings are to be 2 feet wide and 8 inches thick. Determine the cubic yards of concrete needed for the footings.

STEP 1: Determine total length (l) of the footings.

length (l) = 36 + 36 + 24 + 24

l = 120′

STEP 2: Determine total width (w) of footings.

First, multiply the wall width by 2: (8″)(2) = 16″

Since 16 inches is less than 2 feet, the footing width will be 2 feet, the minimum width allowed.

STEP 3: Find cubic feet of concrete using the formula cu. ft. concrete = (l)(W)(0.66)

cu. ft. = (120)(2)(0.66)

cu. ft. = 158.4

STEP 4: Convert cubic feet to cubic yards.

cu. yds. = cu. ft. ÷ 27

cu. yds. = 158.4 ÷ 27

cu. yds. = 5.87

ANSWER: For the purpose of estimating, 6 cubic yards of concrete are needed.

Practical Problems

Solve the problem and round off the answer to the nearest hundredth (two places to the right of the decimal).

1. A building 48 feet long and 30 feet wide is to be constructed using 8 inch blocks. The concrete footings are to be 3 feet wide and 8 inches thick. Determine the cubic yards of concrete needed for the footings. ______________

UNIT 43

Estimating Cubic Yards of Concrete for Slab Work

Basic Principles of Estimating Cubic Yards of Concrete for Slab Work

Concrete floors, patios, and walkways are examples of concrete *slabs*. Residential concrete floors, patios, and walkways are typically placed as 4 inch-thick slabs. Cubic yards of concrete for these 4 inch-thick slabs can be estimated using the following formula.

$$\text{cubic yards (cu. yds.) of concrete} = \frac{(l')(w')(0.33')}{27}$$

When using this mathematical formula, the values for length (l) and width (w) must be represented in units of *feet*. The value of 0.33 feet used in the equation is equivalent to 4 inches, the typical thickness of concrete slabs placed for residential floors, patios, and walkways.

EXAMPLE 1: How many cubic yards of concrete are needed for a residential walkway 32 feet long, 4 feet wide, and 4 inches thick?

STEP 1: Find cubic feet of concrete.

$\text{cubic feet (cu. ft.)} = (l)(w)(0.33)$

$\text{cu. ft.} = (32)(4)(0.33)$

$\text{cu. ft.} = 42.24$

STEP 2: Convert cubic feet to cubic yards.

$\text{cubic yards (cu. yds.)} = \text{cu. ft.} \div 27$

$\text{cu. yds.} = 42.24 \div 27 = 1.56$

ANSWER: For the purpose of estimating, 2 cubic yards of concrete are needed.

Concrete slabs are also used to support freestanding structures such as masonry piers, columns, and chimneys. Slab footings for non-load-bearing piers and columns such as those enhancing driveway entrances may be only 8 inches thick, while slab footings supporting masonry chimneys and load-bearing piers or columns have a thickness of 12 inches or more. Slab footings typically extend at least 6 inches beyond each side of the structure that it supports, making both the length and width of the footing 12 inches more than the length and width of the structure. A slab footing supporting a masonry chimney is shown in the following illustration.

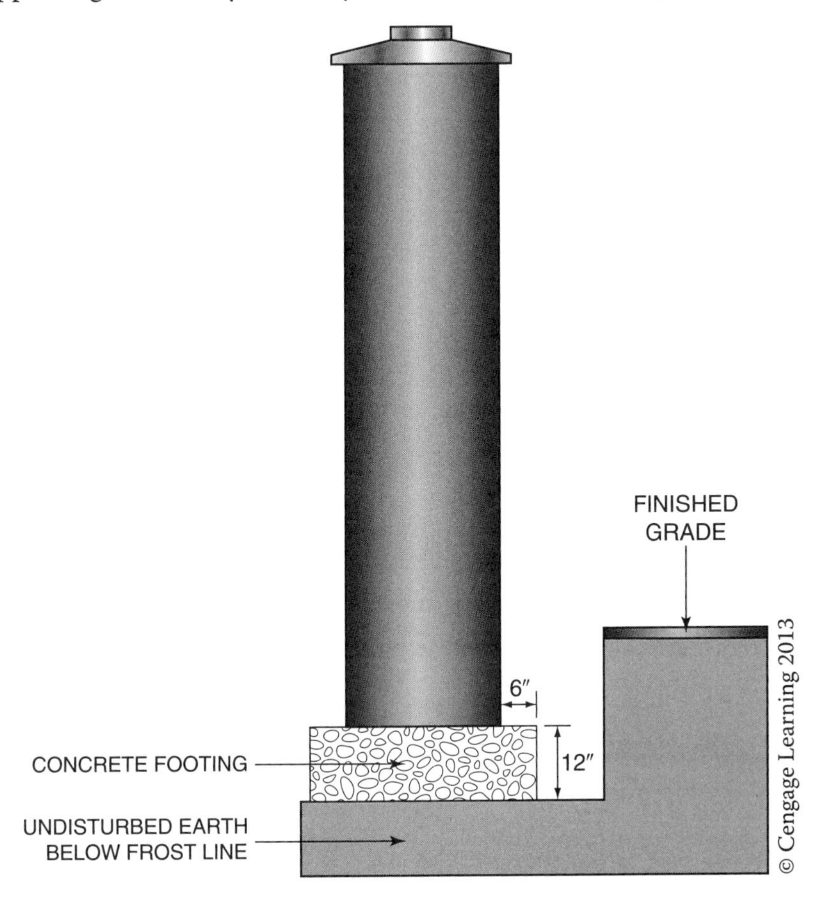

For calculations using the following mathematical formula, values for length (l), width (w), and thickness (t) must be expressed in units of *feet*. Adding a minimum of 6″ beyond each side of the structure the slab footing is intended to support is necessary for deriving the length (l) and width (w) for the footing. Depending on the project, local codes specify how thick (t) the concrete slab footing must be.

EXAMPLE 2: Determine the cubic yards of concrete needed for an 8 inch-thick slab footing supporting a non-load-bearing, brick driveway entrance column whose base dimensions measure 36″ × 42″.

STEP 1: Find slab length by adding 12 inches to each the length and width of the structure.

slab length (l) = 36″ + 12″

slab length = 48″

slab length = 4′

slab width (w) = 42″ + 12″ = 54″

slab width = 4.5′

STEP 2: Find cubic feet of concrete.

cubic feet (cu. ft.) = (l)(w)(0.66′)

cu. ft. = (4′)(4.5′)(0.66′)

cu. ft. = 11.88

STEP 3: Convert cubic feet to cubic yards.

cubic yards (cu. yds.) = cu. ft. ÷ 27

cu. yds. = 11.88 ÷ 27

cu. yds. = 0.44

ANSWER: Approximately 0.44 cubic yards of concrete are needed.

Did you know?

Most suppliers of ready-mixed concrete, a term referring to concrete that is mixed in a truck-mounted mixer while in transit to a worksite, have a minimum delivery policy. Expect extra charges for smaller partial loads.

Practical Problems

Solve each problem and round off all answers to the nearest hundredth (two places to the right of the decimal).

1. A basement floor is 42 feet 6 inches long, 26 feet wide, and 4 inches thick. How many cubic yards of concrete are needed for the floor? __________

2. The concrete floor in a garage is 110 feet 6 inches wide, 162 feet 8 inches long, and 6 inches thick. Find the cubic yards of concrete needed to lay the floor. __________

3. The concrete floor for a factory building is 182 feet 6 inches long, 54 feet 6 inches wide, and 8 inches thick. Find the number of cubic yards of concrete needed for the floor. __________

4. A concrete floor 8 inches thick is laid in a mill building. The inside dimensions of the building are width, 62 feet 8 inches, and length, 137 feet 6 inches. How many cubic yards of concrete are needed for the floor? __________

5. Determine the amount of concrete needed for a floor 120 feet long, 44 feet 9 inches wide, and 4 inches thick. __________

UNIT 44

Applications Involving Estimations of Masonry Materials for Brick and Block Wall Construction

Basic Principles for Applications Involving Estimations of Masonry Materials for Brick and Block Wall Construction

When constructing walls using bricks or concrete blocks, the number of bricks or blocks and the amount of masonry cement and sand each must be estimated prior to beginning construction.

EXAMPLE 1: How many standard-size bricks, bags of masonry cement, and cubic yards of sand are needed to build a single-wythe brick wall whose length is 28 feet and height is 12 feet?

STEP 1: Determine the number of bricks needed.

total wall area (A) = l × w

A = 28′ × 12′

A = 336 square feet

bricks needed = (A in square feet) × (6.75 [feet of brick per square foot for standard-size brick])

brick needed = 336 × 6.75

= 2,268 bricks

STEP 2: Determine the number of 1-cubic-foot–size bags of masonry cement needed.

bags of cement = 2,268 ÷ 125 (quantity of standard-size brick per 1-cubic-foot–size bag of cement)

bags of cement = 18.14

STEP 3: Determine how much sand is needed.

cubic yards of sand = bags of cement ÷ 8

cubic yards of sand = 18.14 ÷ 8

cubic yards of sand = 2.27

ANSWER: This job requires approximately 2,268 bricks, eighteen 1-cubic-foot–size bags of masonry cement, and 2.27 cubic yards of sand.

Practical Problems

Find the estimated amount of materials for each job.

1. Calculate the quantity of standard-size bricks, 1-cubic-foot–size bags of masonry cement, and cubic yards of sand needed to build a single-wythe brick wall whose length is 36 feet and height is 8 feet.

   ❑ the number of standard-size bricks = ________________
   ❑ the number of bags of masonry cement = ________________
   ❑ the number of cubic yards of sand = ________________

2. Calculate the quantity of engineered-size bricks, 1-cubic-foot–size bags of masonry cement, and cubic yards of sand needed to build a single-wythe brick wall whose length is 44 feet 8 inches and height is 10 feet 4 inches.

   ❑ the number of engineered-size bricks = ________________
   ❑ the number of bags of masonry cement = ________________
   ❑ the number of cubic yards of sand = ________________

3. Calculate the quantity of blocks, 1-cubic-foot–size bags of masonry cement, and cubic yards of sand needed to build a block wall whose length is 28 feet and height is 9 feet 4 inches.

   - ❑ the number of blocks = ____________
   - ❑ the number of bags of masonry cement = ____________
   - ❑ the number of cubic yards of sand = ____________

4. Calculate the cubic yards of grout required to fill the hollow cells of a 12 inch block wall whose length is 52 feet and height is 23 feet 4 inches. ____________

UNIT 45

Estimating Materials for Appliance Chimneys

Basic Principles for Estimating Materials for Appliance Chimneys

The primary function of a chimney is to vent combustion by-products to the outside while protecting the structure and its occupants from fire and the life-threatening effects of carbon monoxide. Appliance chimneys are designed to vent burning solid fuels such as wood, wood pellets, coal, natural gas, propane gas, and fuel oil. A typical appliance chimney consists of 2 foot sections of rectangular flue linings made from fireclay (clay capable of withstanding high temperatures without cracking) and joined using medium-duty refractory mortar. Single-wythe brick walls encircle the flue linings. A reinforced concrete chimney cap eliminates moisture penetration at the top of the chimney. The cap should be a minimum of 4 inches thick adjacent to the lining and should slope away from the flue on all sides. The flue lining should extend no less than 2 inches or more than 6 inches above the cap. As shown in the following illustration, the flue lining extends a minimum of 6 inches above the brickwork.

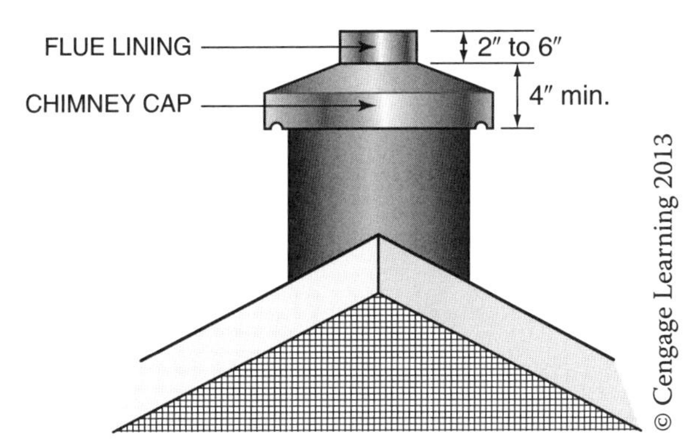

Building code regulations specify the required size of a flue liner based on the intended appliance and its connector. For typical residential appliance chimneys, the nominal sizes of flue linings are 8″ × 8″, 8″ × 12″, and 12″ × 12″. The nominal size refers to the approximate length and width of the opening. The standard length of a single flue lining is 2 feet or 24 inches. The following illustration depicts an appliance chimney using an 8″ × 12″ flue lining enclosed in a single wythe of either standard modular-size or engineered-size brick. Seven standard-size or engineered-size bricks are required for a single course of brickwork. An 8″ × 8″ flue lining requires six standard-size or engineered-size bricks for one course of single-wythe brickwork. A single-wythe of brickwork enclosing a 12″ × 12″ flue lining requires eight standard-size or engineered-size bricks per course. For estimating purposes, the height of a single course of standard-size bricks equals 2¾ inches and the height of a single course of engineered-size bricks equals 3¼ inches. Although it is required that flue linings begin at least 12″ below the lowest inlet, unless specified otherwise, it is to be assumed that the first flue liner rests on the slab footing. The base of the chimney is supported by a concrete slab footing below ground level, its actual depth below the frost line and on soil determined by an engineer to be capable of supporting the chimney's weight. A single round clay thimble is needed to permit connection of the appliance to the chimney.

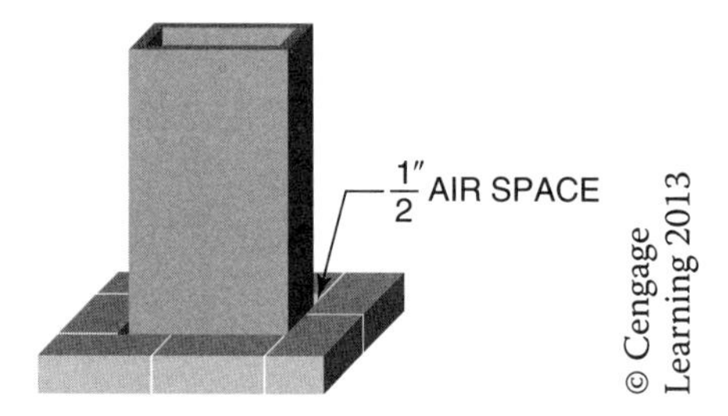

STEP 1: Determine the number of standard-size or engineered-size bricks per course. This is equal to six bricks for an 8″ × 8″ flue lining, seven bricks for an 8″ × 12″ flue lining, and eight bricks for a 12″ × 12″ flue lining.

STEP 2: Determine the overall height of the chimney. The height is taken from the top surface of the slab footing and continues to the top of the last flue lining.

STEP 3: Convert the overall chimney height measurement to *inches*.

STEP 4: Determine the number of courses of bricks. Because the brickwork tops out a minimum of 6 inches below the top of the last flue liner, subtract 6 inches from the overall chimney height. Divide the remaining height in inches by 2¾ if standard-size bricks are specified. Divide the remaining height in inches by 3¼ when engineered-size bricks are specified.

STEP 5: Multiply the number of courses as determined in Step 4 by the number of bricks required for a single course of brickwork as given in Step 1.

STEP 6: Determine the number of flue liners needed. Divide the overall height in *inches* by 24 or the overall height in *feet* by 2.

EXAMPLE 1: How many standard-size bricks are needed to build single-wythe chimney walls for an appliance chimney 24 feet 6 inches tall having an 8″ × 12″ flue lining?

STEP 1: Determine the number of bricks per course.

8″ × 12″ flue lining = 7 bricks per course

STEP 2: Determine the overall height of the chimney.

chimney height = 24′6″

STEP 3: Convert chimney height to inches.

24′6″ = 294″

STEP 4: Determine the number of courses of bricks.

294″ − 6″ = 288″

288″ ÷ 2¾ = 104.7

STEP 5: Number of courses × bricks per course = bricks required

104.7 × 7 = 732.9

ANSWER: Approximately 733 bricks are needed.

EXAMPLE 2: For Example 1, how many sections of 2 foot flue liners are needed to line the chimney from its base to the top?

24′6″ ÷ 2 = 12¼ flue linings

EXAMPLE 3: For Example 1, how many 1-cubic-foot–size bags of masonry cement are needed to lay the brick?

STEP 1: 125 bricks per bag of masonry cement

$733 \div 125 = 5.86$ bags

EXAMPLE 4: For Example 1, how many yards of masonry sand is needed to make the necessary mortar?

cubic yards = number of bags of cement ÷ 8

cubic yards = $5.86 \div 8$

cubic yards = 0.73 yards

Appliance chimneys may have multiple flues. This permits venting more than one appliance, each having its own flue enclosed in a single brick chimney. A 4 inch solid masonry wythe, bonded into the walls of the chimney, is provided between adjacent flues within the chimney walls. Metal wall ties placed every three courses tie the single wythe of brickwork between the flue linings into the exterior chimney walls. Such a chimney is depicted in the following illustration.

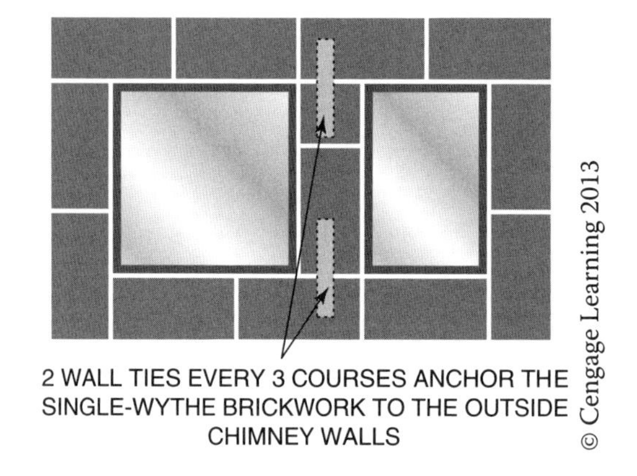

For chimney walls designed as single-wythe brickwork and a single wythe of brickwork between the adjacent flue linings, a single brick course for this chimney having a 12″ × 12″ flue lining, and an 8″ × 12″ flue lining requires 12½ bricks.

Practical Problems

Refer to the following illustration and information to solve problems 1–3. Because answers represent estimates, round up answers having fractions or decimals to the next whole number.

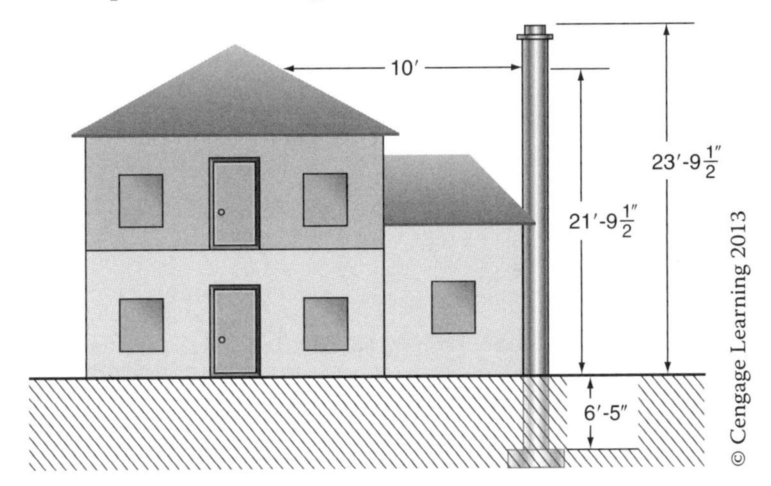

An 8″ × 12″ single-flue brick appliance chimney having single-wythe walls is to be built outside an existing house. The overall height of the chimney from ground level to its top is 23 feet 9½ inches; the minimum height is for it to be at least 2 feet higher than any part of a structure within 10 feet distance. The slab footing must be below the frost line and on soil capable of supporting the chimney. For this job, the top of the footing is 6 feet 5 inches below ground level, also the base of the chimney. Brickwork below ground level is to be constructed as a solid brick pier, requiring an additional three bricks per course as solid fill. 8″ × 12″ flue linings begin at ground level. Engineered-size bricks are chosen for construction.

1. What is the estimate of the number of engineered-size bricks required to build the chimney, including brick fill, from the slab footing to ground level? ____________

2. What is the estimate of the number of bricks required to build the chimney from ground level to the finished height of brickwork if the brickwork is to stop 6 inches below the top of the last flue liner? ____________

3. How many 2 foot sections of flue lining are needed to construct the chimney to its final height if the first flue liner begins at ground level? ____________

4. How many 1-cubic-foot–size bags of masonry cement are needed to lay all bricks, those below grade and those above grade? ____________

5. How many yards of sand are needed to make the mortar for laying the brick? ____________

UNIT 46

Estimating Materials for Masonry Fireplaces

Basic Principles for Estimating Materials for Masonry Fireplaces

The materials required to build a typical fireplace include (1) face bricks for outer walls, surround face, outer hearth, smoke chamber walls, and infill; (2) wall ties; (3) firebricks for the firebox; (4) a reinforced concrete slab for supporting the firebox and hearth extension; (5) a firebox ash dump door and an ash pit cleanout door; (6) combustion air-intake components; (7) a throat damper and insulation; (8) steel lintels; (9) flue liners; (10) a chimney cap; (11) refractory cement; (12) masonry cement; and (13) sand. These materials can be seen in the following cross-sectional view of a fireplace.

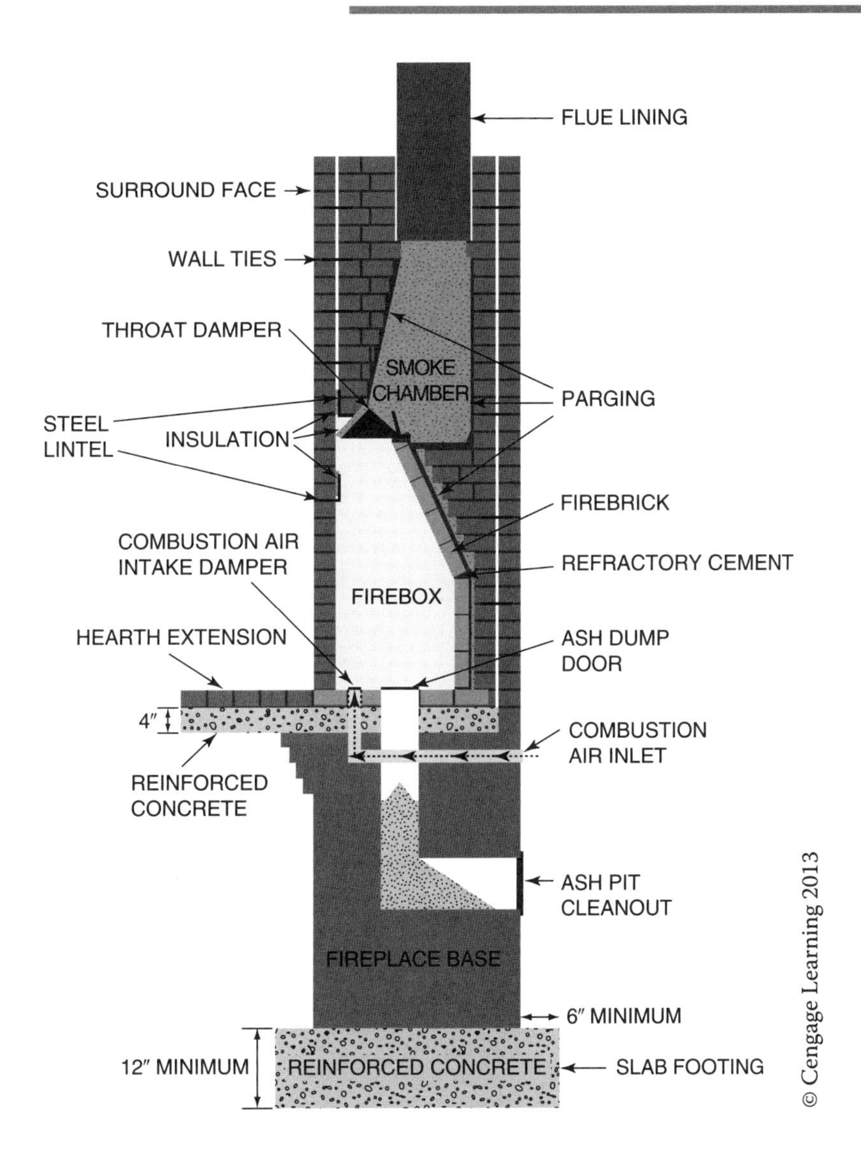

Estimating Materials for Fireplace Base

Fireplace foundation walls should be constructed of *solid masonry*, meaning materials are cored no more than 25%. The walls are to be a minimum of 8 inches thick, having no voids except for the ash pit and an air inlet to the combustion chamber. Masons construct foundation walls using either grout-filled blocks or bricks cored no more than 25%.

The dimensions of the fireplace base depend upon the size of the firebox as well as the outer hearth extension. A fireplace having a firebox opening width of 36 inches could have a base measuring 6 feet long and 2 feet 8 inches wide.

EXAMPLE 1: How many 8 inch blocks are needed to build a fireplace base having a length of 6 feet, a width of 2 feet 8 inches, and a height of 8 feet?

STEP 1: Knowing that the nominal length and width of each block measures 16″ × 8″, find the number of blocks needed for a single course. As the illustration shows, 11 blocks are needed to complete a single course.

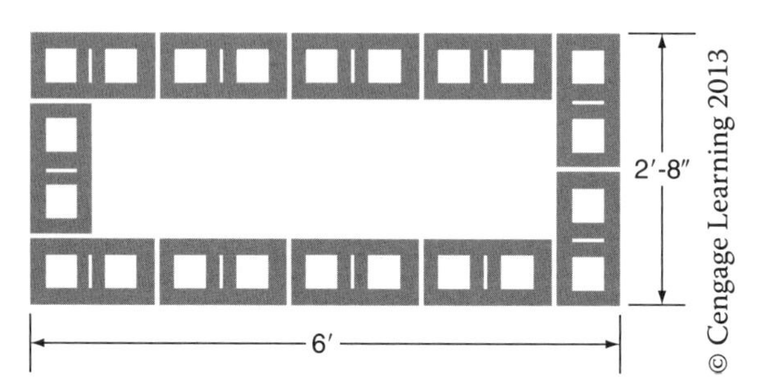

STEP 2: Convert the height of the base to inches.

8′ = (8)(12) = 96″

STEP 3: To determine the number of courses of blocks, divide the base height represented as inches by 8 since one course of block has a height of 8 inches.

96″ ÷ 8 = 12

STEP 4: Multiply the number of blocks per course by the total number of courses.

11 × 12 = 132 blocks

ANSWER: Approximately 132 blocks are needed.

EXAMPLE 2: How many standard-size bricks are needed to build a fireplace base having a length of 6 feet, a width of 2 feet 8 inches, and a height of 4 feet?

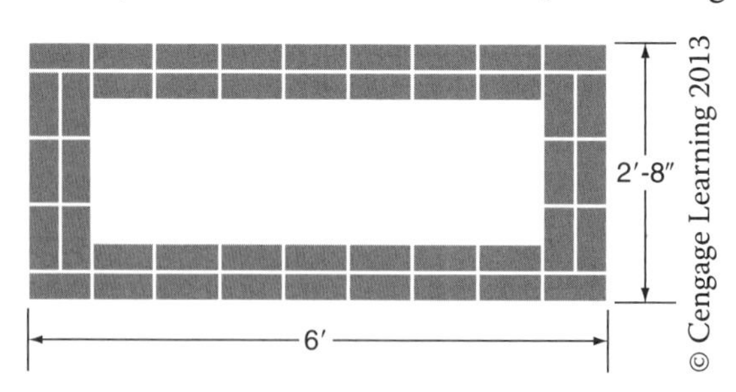

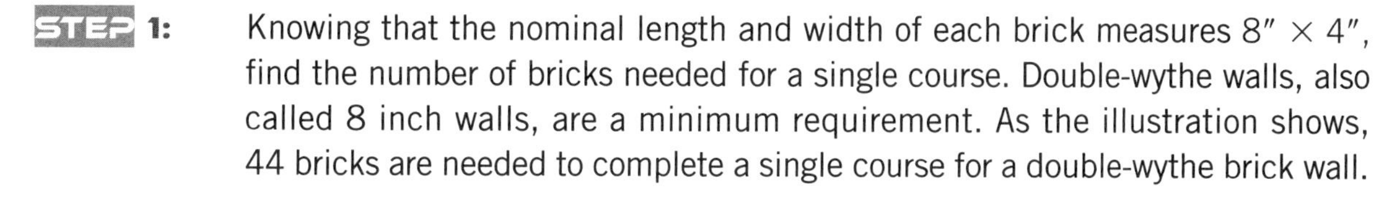

STEP 1: Knowing that the nominal length and width of each brick measures 8″ × 4″, find the number of bricks needed for a single course. Double-wythe walls, also called 8 inch walls, are a minimum requirement. As the illustration shows, 44 bricks are needed to complete a single course for a double-wythe brick wall.

STEP 2: Find the height of the base represented as inches.

$4' = (4)(12) = 48''$

STEP 3: To determine the number of courses of bricks, divide the base height represented as inches by 2¾, since one course of brick has a height of 2¾ inches.

$48'' \div 2\frac{3}{4}'' = 17.45$, rounded up to 18

STEP 4: Multiply the number of bricks per course by the total number of courses.

$(44)(18) = 792$ bricks

ANSWER: Approximately 792 bricks are needed.

Estimating Firebrick for Firebox

Fireboxes contain fires, promote combustion, and radiate heat into room space. They are lined with firebricks that are laid using refractory cement. Firebricks are slightly larger than standard-size bricks, typically measuring 4″ wide, 9″ long, and 2¼″ thick. The following illustration shows the layout for the side and back walls of a firebox having a 36″-wide opening. Firebricks used to build sidewalls and the back wall are often laid in the shiner position, having a bed depth of 2¼″. The opening width determines both the depth and opening height of fireboxes.

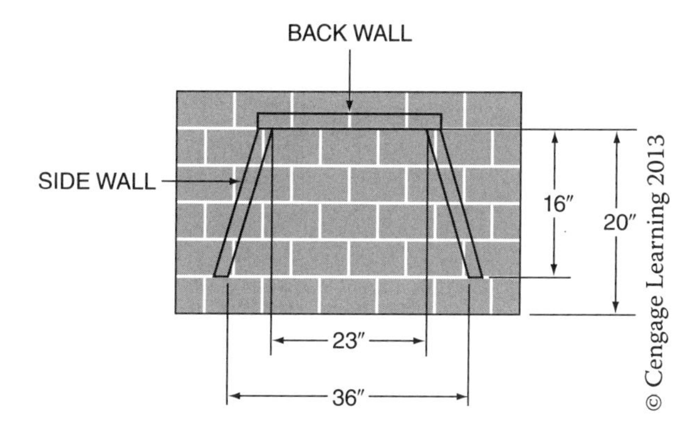

EXAMPLE 1: How many firebricks are needed to construct the floor, also called the inner hearth, for a firebox having a 36″-wide opening?

STEP 1: Find the sum of the opening width and the bed depth of the two firebox sidewalls.

$36'' + 2\frac{1}{4}'' + 2\frac{1}{4}'' = 40\frac{1}{2}''$

STEP 2: Determine the number of whole-length firebricks needed to create the width of the firebox floor, a width allowing the firebrick for the sidewalls to be spaced 36″ apart at the front and laid in the shiner position on top of the firebox floor.

$40\frac{1}{2}'' \div 9 = 4\frac{1}{2}$ firebricks; 5 whole-length firebricks are needed

STEP 3: Determine the number of firebricks needed to create the depth of the firebox floor, a depth allowing the firebrick for the back wall to be spaced 20″ from the front edge of the firebox floor and laid in the shiner position on top of the firebox floor.

The depth of the firebox floor (measured from the face of the fireplace to the back wall of the firebox) is 20″.

The bed depth (thickness of the firebrick forming the back wall of the firebox) is $2\frac{1}{4}''$.

$20'' + 2\frac{1}{4}'' = 22\frac{1}{4}''$

A single row of firebrick including a $\frac{1}{4}''$-wide joint between rows is $4\frac{1}{4}''$.

$22\frac{1}{4}'' \div 4\frac{1}{4}'' = 5\frac{1}{4}$

Because the quotient is more than 5, six full rows of firebrick are needed to create the depth of the firebox

STEP 4: Multiply the number of bricks needed for the width by the number of brick needed for the depth.

$(5)(6) = 30$ firebricks

ANSWER: 30 firebricks are needed for the firebox floor, also called the inner hearth.

EXAMPLE 2: How many firebricks are needed to build the firebox two side walls and the back wall?

STEP 1: Find the sum of the length of the two sidewalls and the back wall.

16″ + 16″ + 23″ = 55″

STEP 2: Divide the sum of the two side walls and back wall by 9, which is the length of 1 firebrick. Round up the answer to the next whole number.

$55 \div 9 = 6\frac{1}{9}$

7 firebricks per course for sidewalls and backwall

STEP 3: Determine the number of courses needed to build the sidewalls and back wall to the specified height. Each course is approximately 4¼″ when laid in the shiner position. The recommended height for the sidewalls and back wall of a firebox having a 36″ opening is 37″, information provided by the Brick Industry Association.

$37'' \div 4\frac{1}{4}'' = 8.7$

9 courses of firebrick

STEP 4: Multiply the number of bricks per course by the number of courses.

(7)(9) = 63 firebricks

ANSWER: Approximately 63 firebricks are needed for sidewalls and back wall.

EXAMPLE 3: How many firebricks are needed to build the entire firebox, the floor, two sidewalls, and back wall?

STEP 1: Find the sum of the firebricks needed for the floor and the firebricks needed for the sidewalls and back wall.

30 + 63 = 93

ANSWER: 93 firebricks are needed.

Estimating Face Brick for the Fireplace Walls, Smoke Chamber Walls, and Chimney Walls

One method used to estimate the number of face bricks needed to build a fireplace relies upon volume calculations. Bedded in mortar, the volume of a standard-size brick is equivalent to approximately 88 cubic inches, and the volume of an engineered-size brick is approximately 104 cubic inches.

$(8'')(4'')(2\frac{3}{4}'') = 88$ cubic inches

Finding the overall volume of a fireplace, one can estimate how many bricks are needed to build it. Subtracting from the overall volume of the fireplace, the combined volume of the firebox, smoke chamber, and flue opening results in the volume that face bricks and mortar occupy.

EXAMPLE 1: A fireplace with a 36 inch-wide opening and a 12″ × 12″ flue is to be built using standard-size bricks. Its overall dimension is 6′ × 2′8″ and its height from the hearth to the top is 16 feet. Estimate the number of bricks needed.

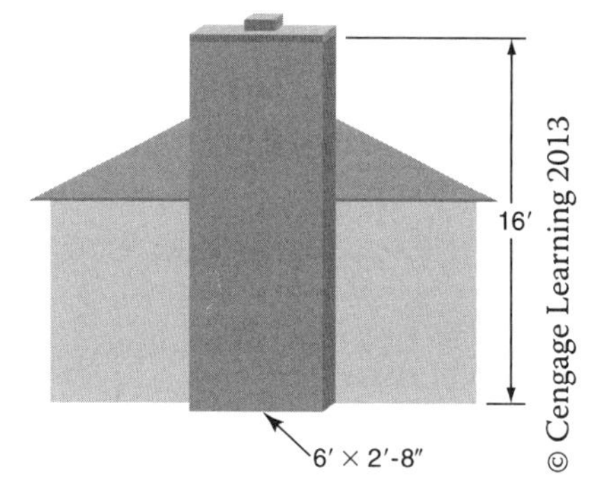

STEP 1: Convert all dimensions to *inches*.

6′ = 72″

2′8″ = 32″

16′ = 192″

STEP 2: Find the total volume (in cubic inches) based on the overall dimensions.

(72″)(32″)(192″) = 442,368 cubic inches

STEP 3: Estimate the volume (in cubic inches) of the firebox.

(36″)(18″)(37″) = 23,976 cubic inches

STEP 4: Estimate the volume (in cubic inches) of the smoke chamber and flue opening. Estimate the length and width of the smoke chamber to be equivalent to the size of the flue opening. Referring to the table and chart provided, a 36 inch-wide firebox is 37 inches tall. The smoke chamber begins above the firebox and the flue opening begins above the smoke chamber.

16′ = 192″

192″ − 37″ = 155″

(12″)(12″)(155″) = 22,320 cubic inches

STEP 5: Subtract the combined volume of the firebox, smoke chamber, and flue opening from the overall volume found in Step 2.

442,368 − 23,976 − 22,320 = 396,072

STEP 6: Divide the answer in Step 5 by 88, which is the volume in cubic inches of one standard-size brick.

396,072 ÷ 88 = 4,500.8

ANSWER: Approximately 4,500 standard-size bricks are needed.

Estimating 2-Foot-Sections of Flue Liners for the Fireplace Chimney

The setting of the first flue liner atop the smoke chamber begins the fireplace chimney. Tables and charts such as those provided by the Brick Industry Association and shown here should be referenced for sizing fireplaces. In so doing, one finds that the recommended opening height for a 36 inch-wide firebox opening is 29 inches and the firebox walls should continue a minimum of 8 inches above the opening, a total of 37 inches and the beginning of the smoke chamber. The recommended height of the smoke chamber, starting at the fireplace throat and ending with the setting of the first section of flue liner, is 27 inches and should not exceed the firebox opening width, 36 inches. The combined height of the firebox assembly, 37 inches, and the recommended height of the smoke chamber, 27 inches, equals 64 inches or 5 feet 4 inches. For the 16 foot-tall fireplace illustrated earlier in this unit, the height of the fireplace chimney is estimated to be approximately 16 feet minus 5 feet 4 inches or 10 feet 8 inches.

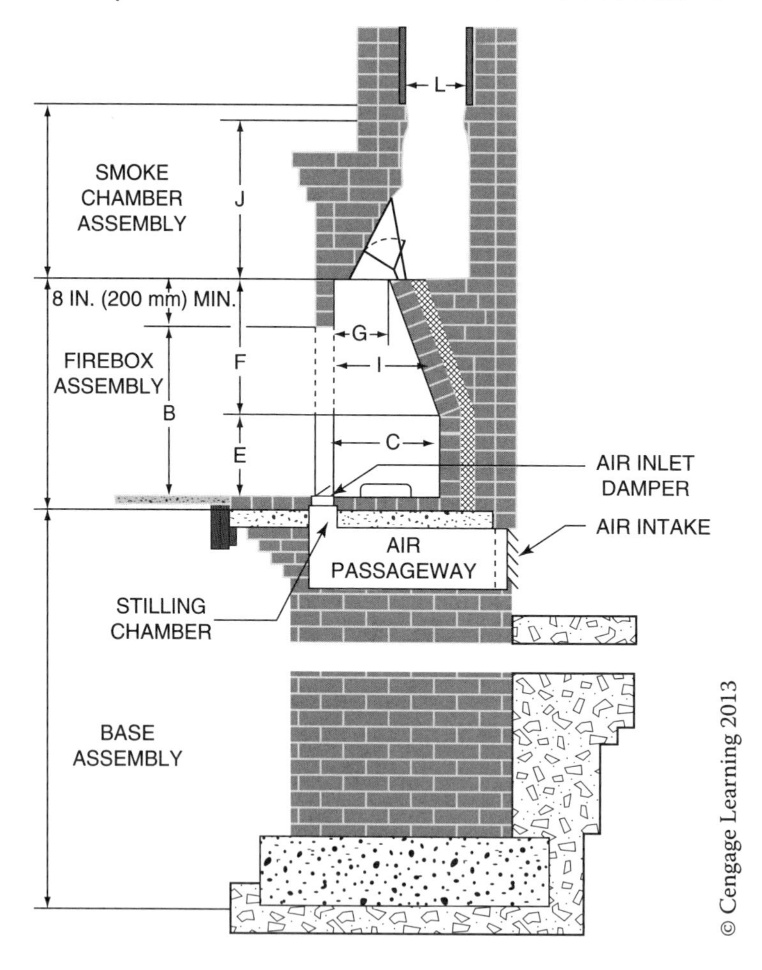

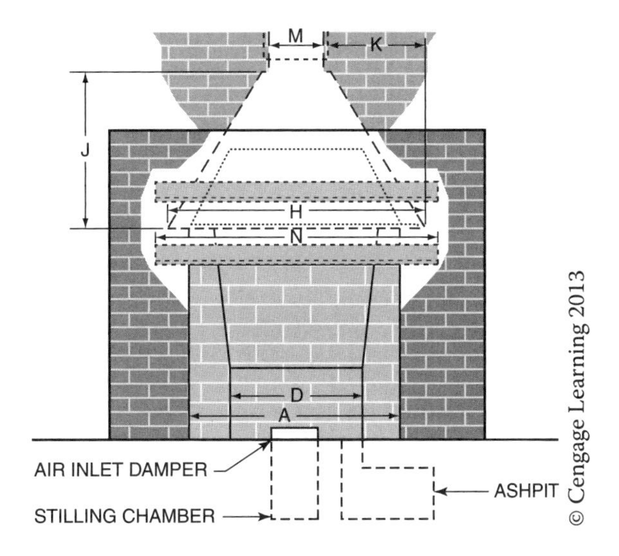

TABLE 1: Single-Face Fireplace Dimensions[a], Inches[b]

	Finished Fireplace Opening						Rough Brick Work[c]				Steel Angle[d]
A	B	C	D	E	F	G	H	I	J	K	N
24	24	16	11	14	18	8¾	32	21	19	10	A-36
26	24	16	13	14	18	8¾	34	21	21	11	A-36
28	24	16	15	14	18	8¾	36	21	21	12	A-36
30	29	16	17	14	23	8¾	38	21	24	13	A-42
32	29	16	19	14	23	8¾	40	21	24	14	A-42
36	29	16	23	14	23	8¾	44	21	27	16	A-48
40	29	16	27	14	23	8¾	48	21	29	16	A-48
42	32	16	29	16	24	8¾	50	21	32	17	B-54
48	32	18	33	16	24	8¾	56	23	37	20	B-60
54	37	20	37	16	29	13	68	25	45	26	B-66
60	37	22	42	16	29	13	72	27	45	26	B-72
60	40	22	42	18	30	13	72	27	45	26	B-72
72	40	22	54	18	30	13	84	27	56	32	C-84

[a]Adapted from *Book of Successful Fireplaces,* 20th Edition.
[b]SI conversion: mm = in. × 25.4.
[c]L and M, shown in the previous figures, are equal to outside dimensions of flue lining plus at least 1 inch (25 mm). When determining flue lining dimensions, note that L should be greater than or equal to M.
[d]Angle sizes: A—3 × 3 × ¼ inches, B—3½ × 3 × ¼ inches, C—5 × 3½ × 5⁄16 inches

Estimating the Fireplace Surround

The fireplace surround is the face of the fireplace immediately surrounding the firebox opening. Typical materials used for its construction include stones, bricks, tiles, marble, or a combination of these materials. A steel lintel supports brickwork above the opening, and corrosion-resistant metal ties embedded in the mortar joints of the brickwork behind the fireplace surround anchor the fireplace surround to the rest of the fireplace.

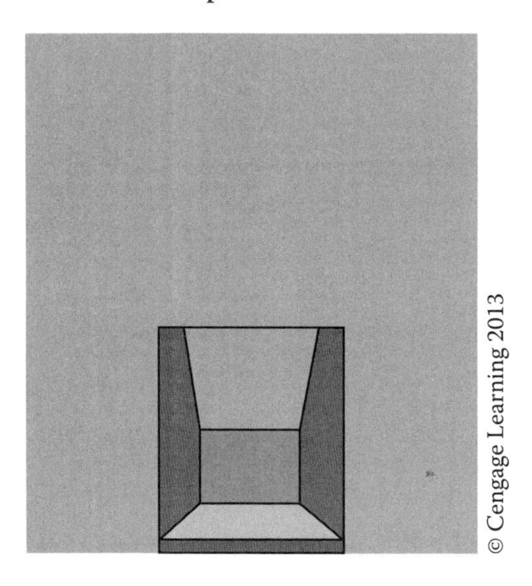

EXAMPLE 1: How many standard-size bricks are needed to construct a fireplace surround whose width is 7 feet and height is 8 feet? The firebox opening measures 36 inches wide and 29 inches tall.

STEP 1: Find the overall area of the fireplace surround.

$(7')(8') = 56$ square feet

STEP 2: Find the area of the firebox opening in units of square inches.
$(36'')(29'') = 1{,}044$ square inches

STEP 3: Convert square inches to square feet. (Dividing square inches by 144 converts square inches to square feet.)

$1044 \div 144 = 7.25$ square feet

STEP 4: Subtract the opening size area from the overall area.

$56 - 7.25 = 48.75$ square feet

STEP 5: Multiply the remaining area found in Step 4 by 6.75, the number of standard-size bricks per square foot.

$(48.75)(6.75) = 329$ standard-size bricks

ANSWER: 329 standard-size bricks are needed.

Estimating Bricks for the Outer Hearth

The outer hearth or hearth extension extends beyond the fireplace opening and into the room. For openings less than 6 square feet it must extend a minimum of 8 inches beyond each side of the opening and 16 inches in front of the opening. For openings 6 square feet or larger, the outer hearth must extend at least 12 inches beyond each side and 20 inches in front of the opening. A raised hearth extension is usually constructed when the level of the firebox floor is higher than the room floor, typically keeping the outer hearth and firebox floor at the same level. Although a raised hearth extension reduces floor space, it offers inviting fireside seating.

A single course of brick is all that is needed when outer hearths are designed to be at floor level, providing the reinforced concrete slab support is placed properly.

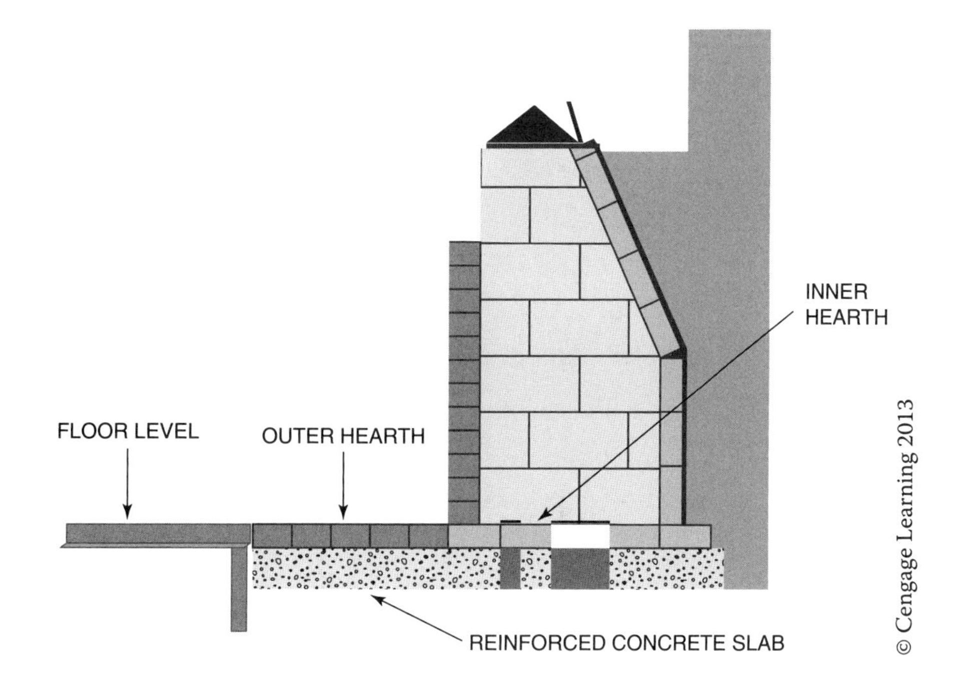

When face bricks are the selected material for outer hearths, they are laid in either the stretcher position or the rowlock position. Laid in the stretcher position with the bottom side bedded in mortar, each brick creates 32 square inches of surface area. This is equivalent to 4½ bricks per square foot.

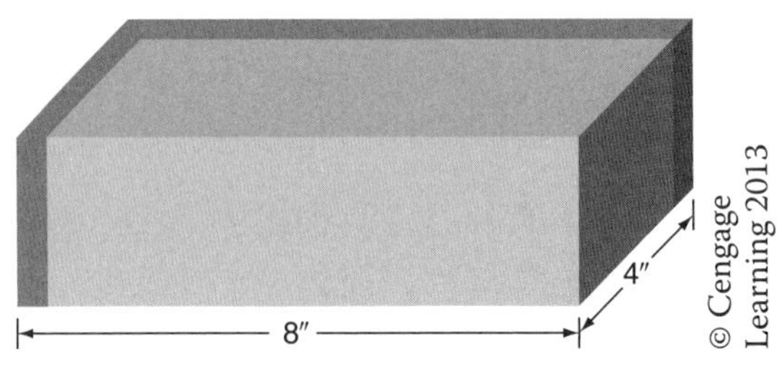

Stretcher position: approximately 4½ standard-size or engineered-size bricks per square foot.

Laid in the rowlock position with the bottom side bedded in mortar, each standard-size brick creates 22 square inches of surface area. There are an estimated 6¾ bricks per square foot for the surfaces of outer hearths constructed with standard-size bricks laid in the rowlock position. It takes an estimated 5½ bricks per square foot for hearths using engineered-size bricks and laid in the rowlock position.

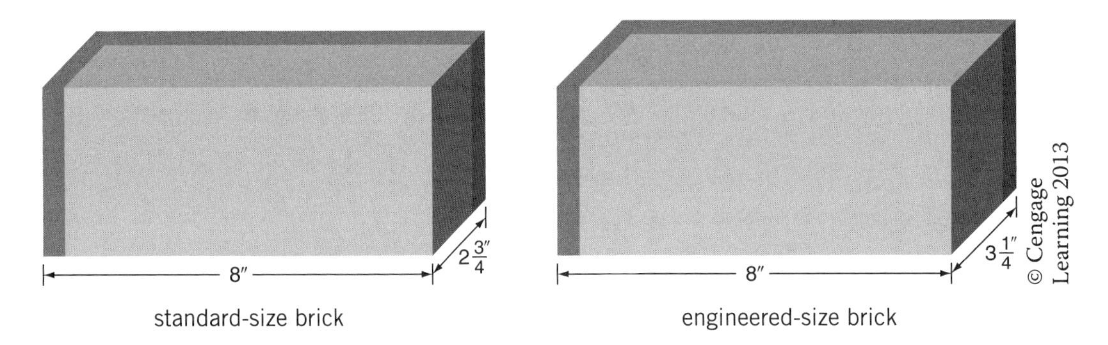

Rowlock position: approximately 6¾ standard-size or 5½ engineered-size bricks per square foot.

EXAMPLE 1: Estimate the number of standard-size solid bricks needed to construct a fireplace outer hearth 8 feet wide and 20 inches deep. Bricks are laid as brick pavers, requiring 4½ bricks per square foot.

STEP 1: Find the area of the outer hearth represented as square feet.

$(8')(1.66') = 13.28$ square feet

STEP 2: Multiply the number of square feet by 4.5, which is the number of standard-size or engineered-size bricks per square feet when laid as brick pavers (the bottom side of the brick bedded in mortar).

$(13.28)(4.5) = 59.76$ standard-size bricks, rounded to 60

ANSWER: Approximately 60 standard-size bricks are needed.

For raised hearths, first calculate the number of bricks required for a single course using the previously explained method. Then multiply the bricks required for a single course by the number of courses of brickwork required for the specified height.

Practical Problems

Solve each problem and round off all answers to the nearest hundredth (two places to the right of the decimal).

Find estimates of the materials needed to build a fireplace as shown in the following illustrations.

1. What is the estimated number of standard-size bricks needed to build the fireplace shown in the following illustration? ____________

   The fireplace calls for a single-face firebox having a 36 inch-wide opening and connected to a 12″ × 12″ chimney flue. The combined height of the firebox and smoke chamber is 5 feet 6 inches. Flue liners begin 5 feet 6 inches above the hearth and extend 6 inches beyond the top of the brickwork.

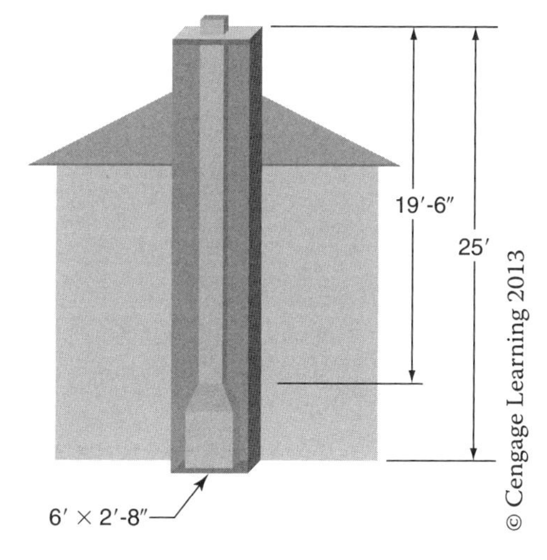

2. What is the estimated number of standard-size bricks needed for the fireplace surround? ______________

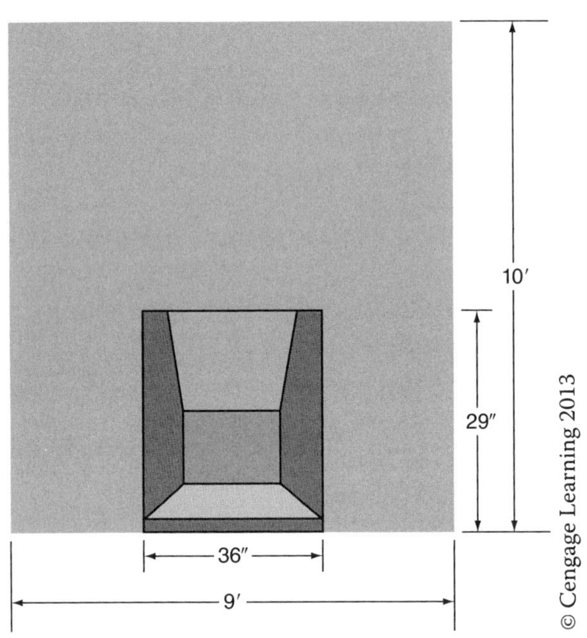

3. What is the estimated number of standard-size bricks needed to construct an outer hearth extending the full width of the fireplace surround in the previous illustration and 20 inches from its face? A single course of bricks laid in the rowlock position creates the hearth. ______________

4. The following chimney is 29 feet 4 inches high. There are six standard-size bricks per course. How many bricks are needed for the entire chimney? __________

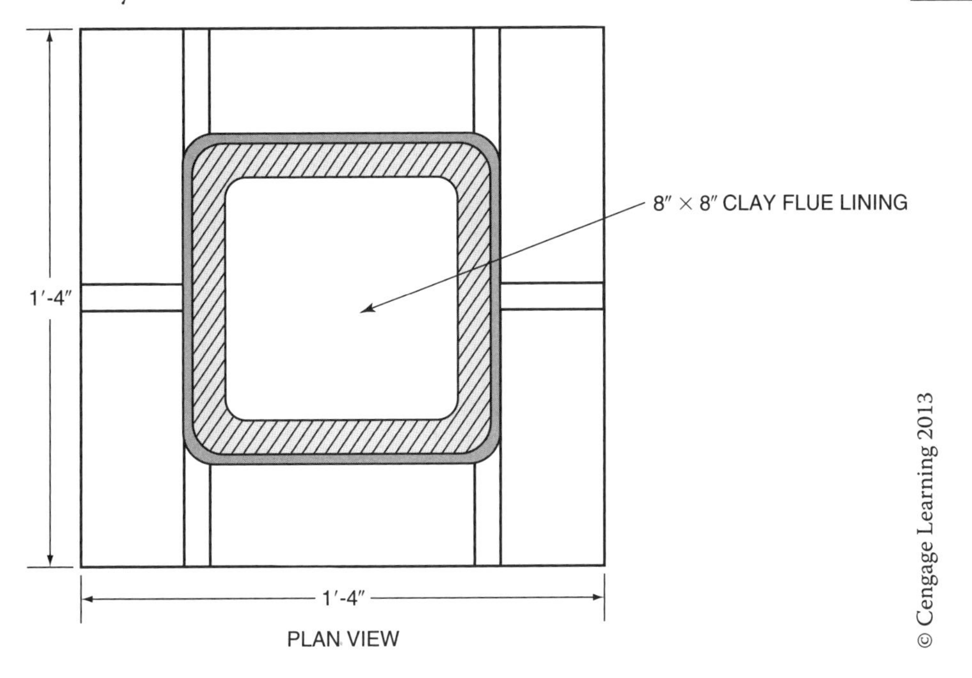

5. If the chimney shown in problem 4 is 36 inches high, and there are six standard-size bricks per course, how many bricks are required for the entire chimney? __________

UNIT 47

Estimating Materials for Brick Paving

Basic Principles for Estimating Materials for Brick Paving

Because of its permanence, low maintenance, and beauty, mortared brick paving is a popular alternative to exposed concrete for walkways, porches, stoops, patios, pool decking, and driveways. Bricks intended for these applications are known as *paving bricks* or *brick pavers*. Rated according to their resistance to weathering, density, slip resistance, and skid resistance, paving bricks outperform face bricks when used for paving. As the name implies, mortared brick paving involves bedding bricks in mortar. Mortar bonds the brick to a rigid base, typically a 4 inch-thick reinforced concrete slab. Modular paving brick, measuring 3⅝ inches wide and 7⅝ inches long, allows for bond pattern uniformity when spaced ⅜ inches apart. For outdoors, type M mortars are recommended for bedding the bricks and filling the mortar joints between them, a process referred to as *grouting the joints*. Mortar is mixed using the same proportions of ingredients as used for usual bricklaying. A 1-cubic-foot–size bag of masonry cement makes a batch of mortar sufficient to bed and grout the joints of approximately 100 paving bricks.

The surface of typical brick pavers measures either 3⅝″ × 7⅝″ or 4″ × 8″. With ⅜ inch-wide mortar joints between them, a single modular-size paving brick is equivalent to 32 square inches. It takes 4½ bricks to cover 1 square foot of surface area.

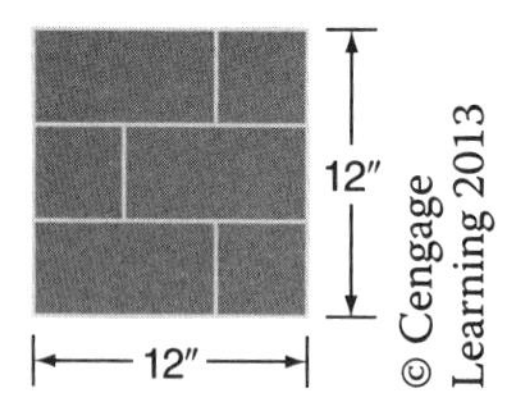

Four popular brick patterns used for brick paving are shown here. The actual quantity of brick needed for a given job may be slightly more depending on the pattern selected.

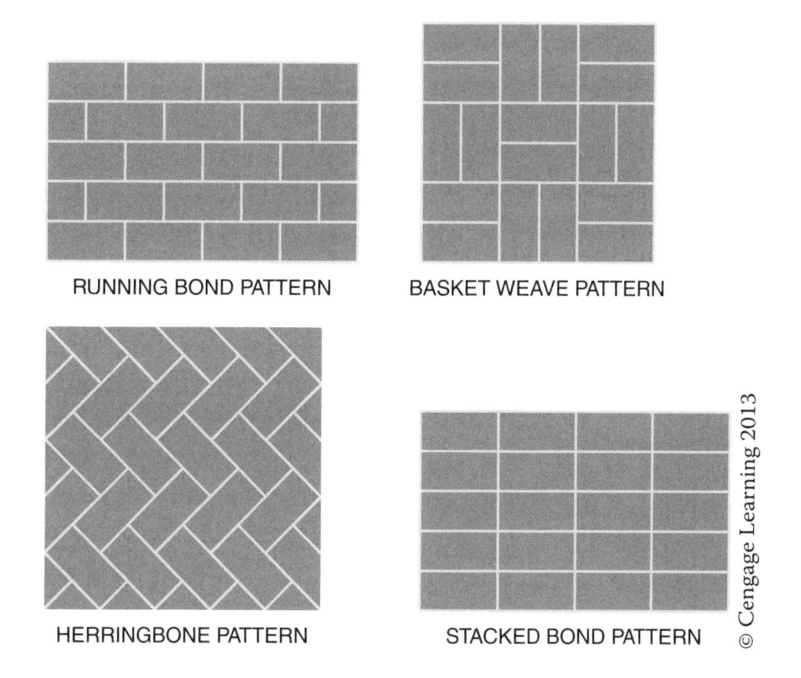

EXAMPLE 1: How many brick pavers are needed to complete a sidewalk that measures 3 feet wide and 32 feet long?

STEP 1: Calculate surface area using the formula area (A) = lw.

A = (3′)(32′)

A = 96 square feet

STEP 2: Multiply the surface area (square feet) by 4.5.

number of brick pavers = (sq. ft.)(4.5)

number of brick pavers = (96)(4.5)

number of brick pavers = 432

ANSWER: 432 brick pavers are needed for this sidewalk.

EXAMPLE 2: How many paving bricks are needed to cover a concrete patio having a length of 20 feet and a width of 15 feet?

STEP 1: Determine the area of the patio using the formula area (A) = lw.

A = (20′)(15′)

A = 300 square feet

STEP 2: Multiply the number of square feet by 4.5.

number of brick = (a)(4.5)

number of brick = (300)(4.5)

number of brick = 1,350

ANSWER: 1,350 paving bricks are needed.

EXAMPLE 3: Assuming that 100 paving bricks can be laid with each 1-cubic-foot–size bag of masonry cement, how many 1-cubic-foot–size bags of Type M masonry cement are needed to bed 1,350 paving bricks and grout the joints between them?

STEP 1: Divide the total number of paving bricks by 100.

number of bags = the number of bricks ÷ 100

number of bags = 1350 ÷ 100

number of bags = 13.5

ANSWER: 13.5 bags of cement are needed.

Practical Problems

Solve each problem and round off all answers to the nearest hundredth (two places to the right of the decimal).

1. How many brick pavers are needed to pave a driveway measuring 15 feet wide by 32 feet long? ______

2. Using the basket weave pattern, how many pavers are needed for a patio that is 28′ × 12′6″? ______

3. How many brick pavers are needed to complete decking around a hot tub if the total patio measures 22 feet in length and 16 feet in width, and the area of the hot tub is 50.24 square feet? ______

4. How many brick pavers are needed for a patio measuring 12′6″ × 18′8″? ______

5. How many brick pavers and how many 1-cubic-foot–size bags of Type M masonry cement are needed to complete the patio and walkway as shown in the following illustration?

   brick pavers ______

   bags Type M masonry cement ______

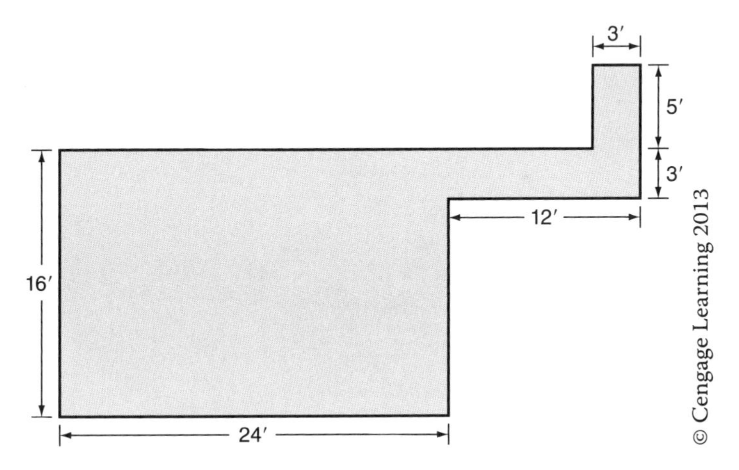

UNIT 48

Estimating Materials for Brick Stairs

Basic Principles of Estimating Materials for Brick Stairs

For the same reasons that bricks are used for walkways and patios—theirs permanence, low maintenance, and beauty—bricks are used to construct stairs. A brick's dimensions simplify the design for a step. Additionally, brick stairs can typically be constructed using the same bricks as used on the house façade.

Stairs have two elements, the *riser* and the *tread*. The riser is the vertical component and the *rise* is the vertical height between two adjoining horizontal treads. The tread is the surface stepped upon. The *tread depth* is measured from the tread's front edge to its back edge. The following illustration identifies the parts of a stair.

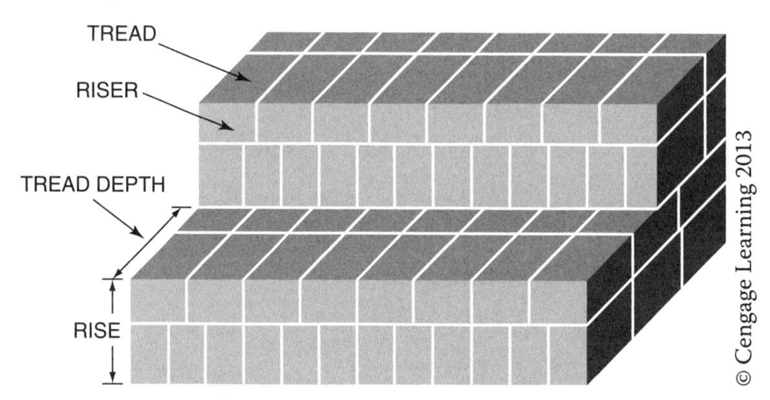

Forming a stair with two courses of bricks, one laid in the stretcher position and the other in the rowlock position, creates a proper rise, approximately 6¾ inches, with standard-size bricks, or 7 inches with engineered-size bricks. As shown in these two illustrations, the actual tread surface can be either a stretcher course or rowlock course of bricks. Choosing to make the tread depth equivalent to the length of 1½ bricks results in a tread depth between 11½ inches and 13 inches, depending on the actual length of the selected bricks and the width of the mortar joint between the front and back rows of rowlocks.

EXAMPLE 1: How many standard-size bricks are needed to construct two stairs, each having a nominal tread depth of 12 inches and a nominal tread width of 32 inches? Each stair consists of two courses of bricks, the first being a stretcher course and the second a rowlock course.

First, find the number of bricks required for the first course of brickwork.

STEP 1: Multiply the specified number of stairs by 12 inches, which is the given tread depth for each stair.

(2)(12) = 24″

STEP 2: Find the area of the first course of brickwork by multiplying the answer found in Step 1 by 32 inches, which is the specified tread width (length of a stair).

(24″)(32″) = 768 square inches

STEP 3: Find the number of standard-size bricks needed for the first course of the first stair (brick laid in the stretcher position) by dividing the answer in Step 2 by 32, which is the number of square inches of a single brick bedded in the stretcher position.

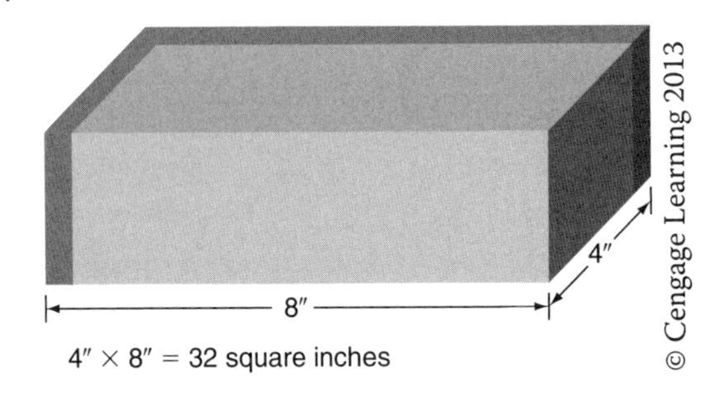

768 ÷ 32 = 24 bricks

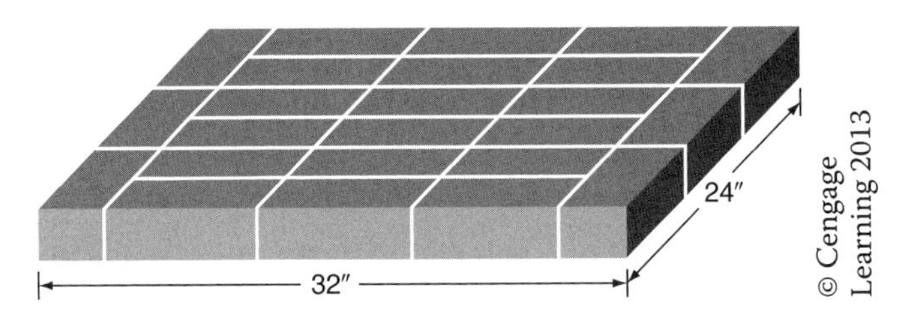

STEP 4: Now find the number of bricks needed for the second course of the first stair (brick laid in the rowlock position) by dividing the answer in Step 2 by 21.4, which is the number of square inches represented by the face side of a single brick laid in a rowlock position.

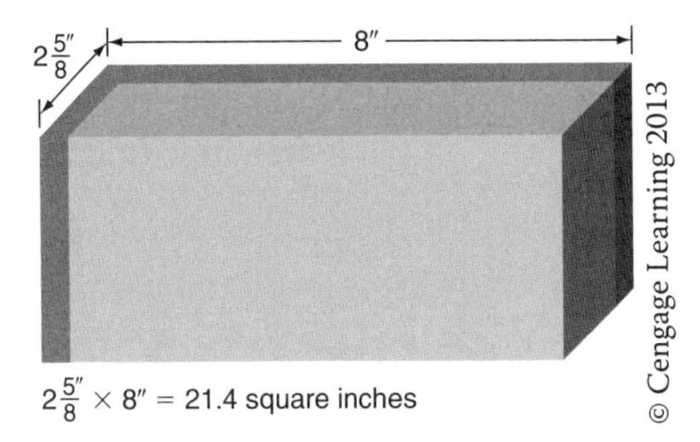

768 ÷ 21.4 = 35.88, rounded up to 36 bricks

STEP 5: Find the bricks needed for the combined two courses of the first stair by adding together the answers found in Steps 3 and 4.

24 + 36 = 60 bricks

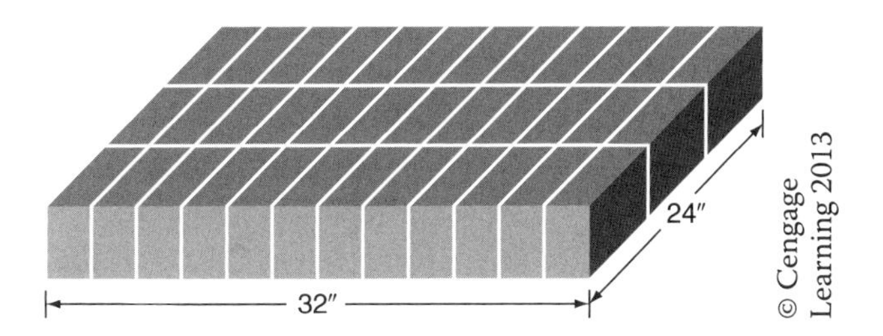

STEP 6: Divide the answer found in Step 5 by 2 to determine the number of bricks required for the second stair.

60 ÷ 2 = 30 bricks

STEP 7: Add the number of bricks found in Steps 5 and 6 to obtain an estimate of the total amount of bricks needed for both steps.

60 + 30 = 90 bricks

ANSWER: The total number of bricks needed for the stairs is 90.

Note: For multistep construction exceeding two stairs, perform Steps 1–5 as in this example to find the number of bricks required for the two courses of bricks making up the first stair. Then find the number of bricks needed for each additional stair by dividing that number by the fractional representation of each additional stair and adding these numbers together to find the total number of bricks needed for all of the stairs.

For example, if 120 bricks are needed to construct the first of four stairs, then:

- (120)(¾) = 90, the number of bricks needed for the second stair
- (120)(²⁄₄) = 60, the number of bricks needed for the third stair
- (120)(¼) = 30, the number of bricks needed for the top stair
- 120 + 90 + 60 + 30 = 300 bricks needed to construct all four stairs

Alternate method for solving this example:

STEP 1: Divide the specified tread width (length of a stair) by 8, which is the nominal length of a single brick, to determine the number of bricks needed to create the tread width.

$32'' \div 8 = 4$ brick (tread width)

STEP 2: Divide the depth of a single tread (inches) by 4, which is the nominal width (inches) of a single brick.

$12 \div 4 = 3$

STEP 3: Multiply the specified number of stairs (in this example, two) by the answer found in Step 2 to find the number of rows (or 4 inch wythes) of bricks needed for the first course.

$(2)(3) = 6$ bricks (combined tread depth for two steps)

STEP 4: Multiply the answer found in Step 1 by the answer found in Step 3.

$(4)(6) = 24$ bricks (number of bricks for the first course, a stretcher course)

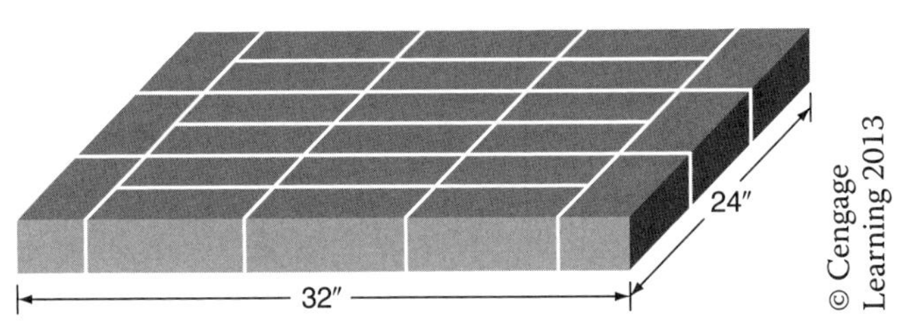

STEP 5: To find the number of bricks required for the second course of brickwork, a rowlock course, multiply the answer in Step 1 by 3, which is the number of standard-size bricks laid in the rowlock position equivalent to the length of one brick laid in the stretcher position.

$(4)(3) = 12$ bricks

STEP 6: Divide the tread depth of a single stair (inches) by 8, the length (inches) of a standard-size brick laid in the rowlock position, to determine the number of rows of bricks laid in the rowlock position creating a single tread.

$12 \div 8 = 1.5$

STEP 7: Multiply the answer in Step 6 by the total number of treads. *Note:* Do not round up the answer in Step 6. Leave it as a decimal or fraction.

$(1.5)(2) = 3$

STEP 8: Multiply the answer in Step 5 by the answer in Step 7 to find the number of bricks needed for the second course of brickwork.

$(12)(3) = 36$ bricks

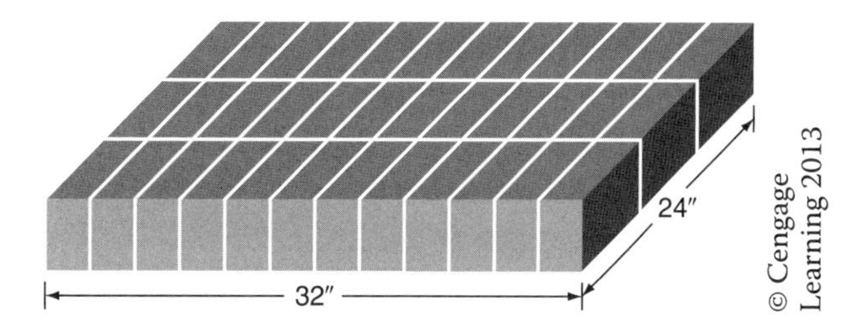

STEP 9: Add together the number of bricks in Steps 4 and 8 to find the total number of bricks needed for the first two courses of brickwork.

$24 + 36 = 60$ bricks

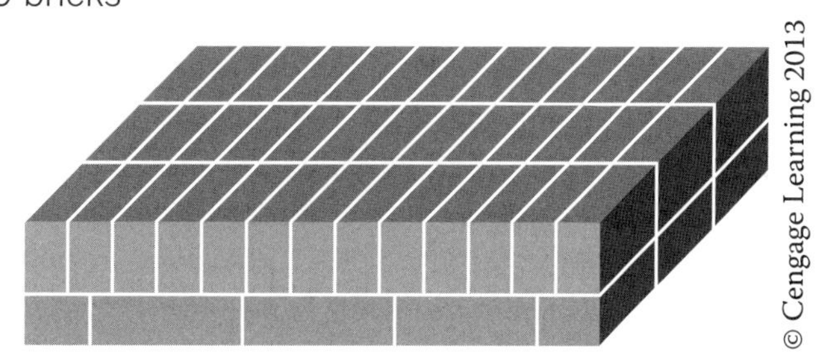

STEP 10: Divide the number of bricks in Step 9 by 2 to find the number of bricks needed for the second stair.

$60 \div 2 = 30$ bricks

STEP 11: Add together the number of bricks in Steps 9 and 10 to find the total number of bricks needed for the two stairs.

60 + 30 = 90 bricks

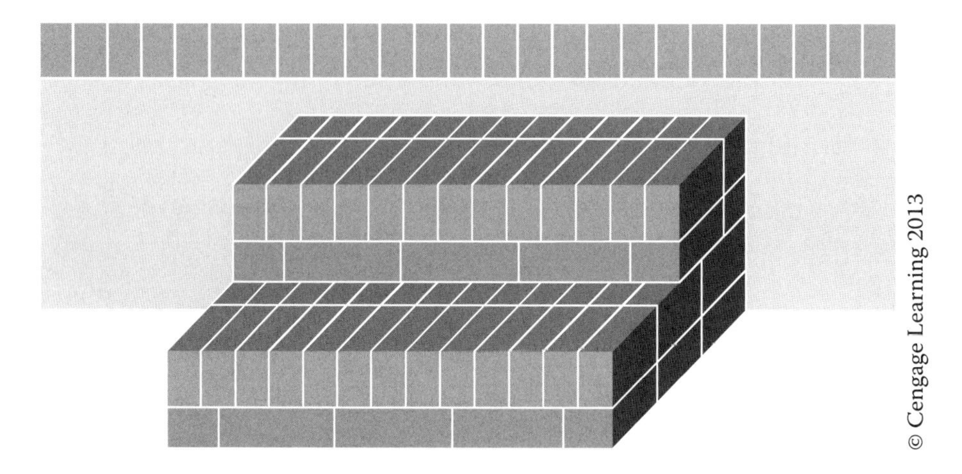

Note: For multistep construction exceeding two stairs, perform procedural Steps 1–9 to find the number of bricks required for the two courses of bricks making up the first stair. Then find the number of bricks needed for each additional stair by dividing that number by the fractional representation of each additional stair as shown in the previous method.

Another method for making brick stairs involves constructing concrete stairways first. The selected brick's dimensions are used as modular measurements for the width and depth of the treads as well as the height of the risers while conforming to building code standards. This consideration eliminates unnecessary cutting of bricks, giving stairs a more uniform appearance. The following is an illustration of a concrete stairway with landing platforms at the top and bottom.

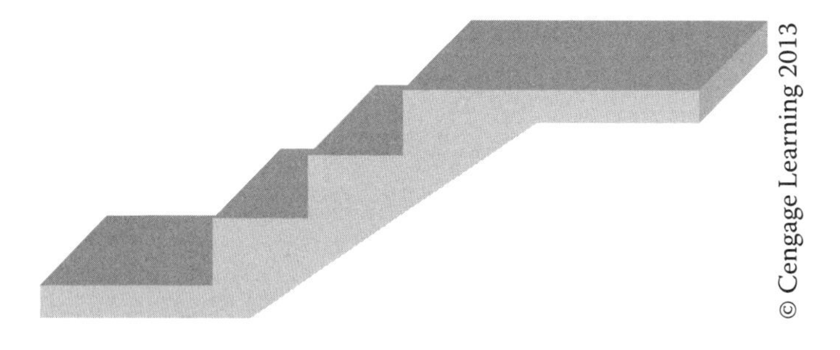

Forming and placing concrete is a task typically performed by carpenters or concrete finishers. But masons may be relied upon to provide these workers with the depth and width of the treads and height of the risers based on the dimensions of the selected brick and intended mortar joint widths.

A single course of standard-size solid brick paves each of the concrete tread surfaces. Choosing to make the tread depth equivalent to the length of 1½ bricks results in a tread depth between 11½ inches and 13 inches, depending on the actual length of the selected brick and the width of the mortar joint adjoining a full-length brick to a half-length brick. Risers consist of three courses of standard-size brick between each brick tread with the exception of the first riser being only two courses of stretchers.

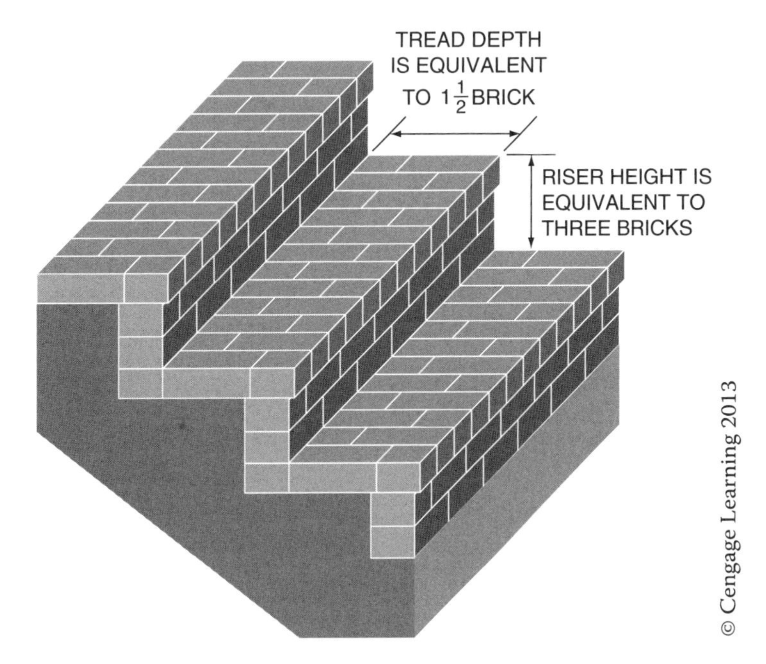

EXAMPLE 2: How many standard-size bricks are needed to cover the riser and tread surfaces of a flight of 3 concrete stairs if each tread measures 12 inches deep by 48 inches wide and each riser is 8 inches high?

STEP 1: Find the area of a single tread using the formula area (A) = length (l) × width (w)

$A = 4 \times 1$

$A = 4$ square feet

STEP 2: Multiply the area of a single tread by the number of treads to get the total area of all treads.

$4 \times 3 = 12$ square feet

STEP 3: Calculate the number of bricks needed for all three treads. Multiply the total area by 4.5, which is the number of bricks per square foot.

$12 \times 4.5 = 54$ bricks

STEP 4: Calculate the area of a single riser.

$4 \times 0.66 = 2.64$ square feet

STEP 5: Calculate the combined area of all risers.

$2.64 \times 3 = 7.92$ square feet

STEP 6: Estimate the number of bricks needed for all risers. Multiply the total square feet of all risers by 6.5, which is the number of bricks per square foot.

$8 \times 6.75 = 53.46$ bricks, rounded up to 54 bricks

STEP 7: Find the sum of the bricks for the treads and the bricks for the risers.

$54 + 54 = 108$ bricks

ANSWER: 108 bricks are needed.

Laying bricks in the rowlock position to serve as tread surfaces and one course of bricks laid in the stretcher position below the rowlock is a popular method for resurfacing concrete stairways with bricks. This technique allows masons to create shorter risers than the method described in the previous illustrated example. The following illustration shows such a flight of stairs. Risers are approximately 6¾ inches when using standard-size bricks or 7 inches with engineered-size bricks, a full inch or less than the riser height created in the previous method. Choosing to make the tread depth equivalent to the length of 1½ bricks results in a tread depth between 11½ and

13 inches, depending on the actual length of the selected brick and the width of the mortar joint between the adjoining bricks.

EXAMPLE 3: Using a stretcher course and a rowlock course to form each stair, how many standard-size bricks are needed to face the surfaces of a flight of three concrete stairs if each tread measures 12″ deep by 40″ wide?

STEP 1: Find the area of a 16 inch rowlock whose length is equal to the width of the stairs. Set up the problem using the formula area (A) = length (l) × width (w)

$A = 3.33' \times 1.33$

$A = 4.43$ square feet

STEP 2: Multiply the area of a single 16 inch rowlock course by the number of treads to get the total area of all rowlocks.

$4.43 \times 3 = 13.29$ square feet

STEP 3: Calculate the number of bricks needed for all three 16 inch rowlocks. Multiply the total area by 6.75, which is the number of bricks per square foot when laid in a rowlock position.

$13.29 \times 6.75 = 89.71$ bricks, rounded up to 90 bricks

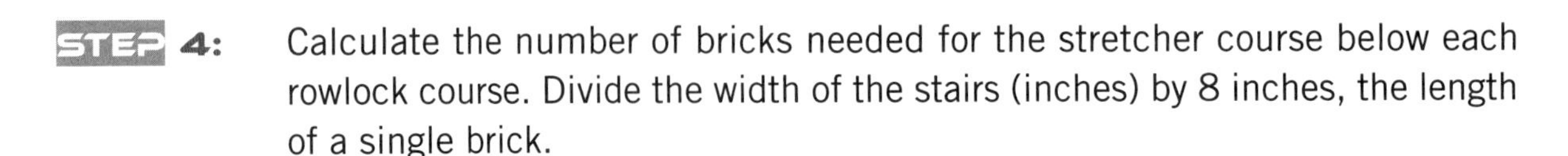

STEP 4: Calculate the number of bricks needed for the stretcher course below each rowlock course. Divide the width of the stairs (inches) by 8 inches, the length of a single brick.

$40'' \div 8 = 5$ bricks

STEP 5: Calculate the number of bricks laid in the stretcher position for the combined number of stairs.

$5 \times 3 = 15$ bricks

STEP 6: Find the sum of the bricks needed for the rowlock courses and the stretcher courses.

$90 + 15 = 105$ bricks

ANSWER: 105 bricks are needed.

Practical Problems

Solve each problem and round off all answers to the nearest hundredth (two places to the right of the decimal).

1. Determine how many standard-size bricks are needed to construct three stairs, each having a nominal tread depth of 12 inches and a nominal tread width of 48 inches. Each stair consists of two courses of bricks, the first being a stretcher course and the second a rowlock course. ____________

2. Using a stretcher course and a rowlock course to form each stair, how many standard-size bricks are needed to face the riser and tread surfaces of a flight of four concrete stairs if each tread measures 12″ deep by 36″ wide? ____________

Use this illustration to answer questions 3–5.

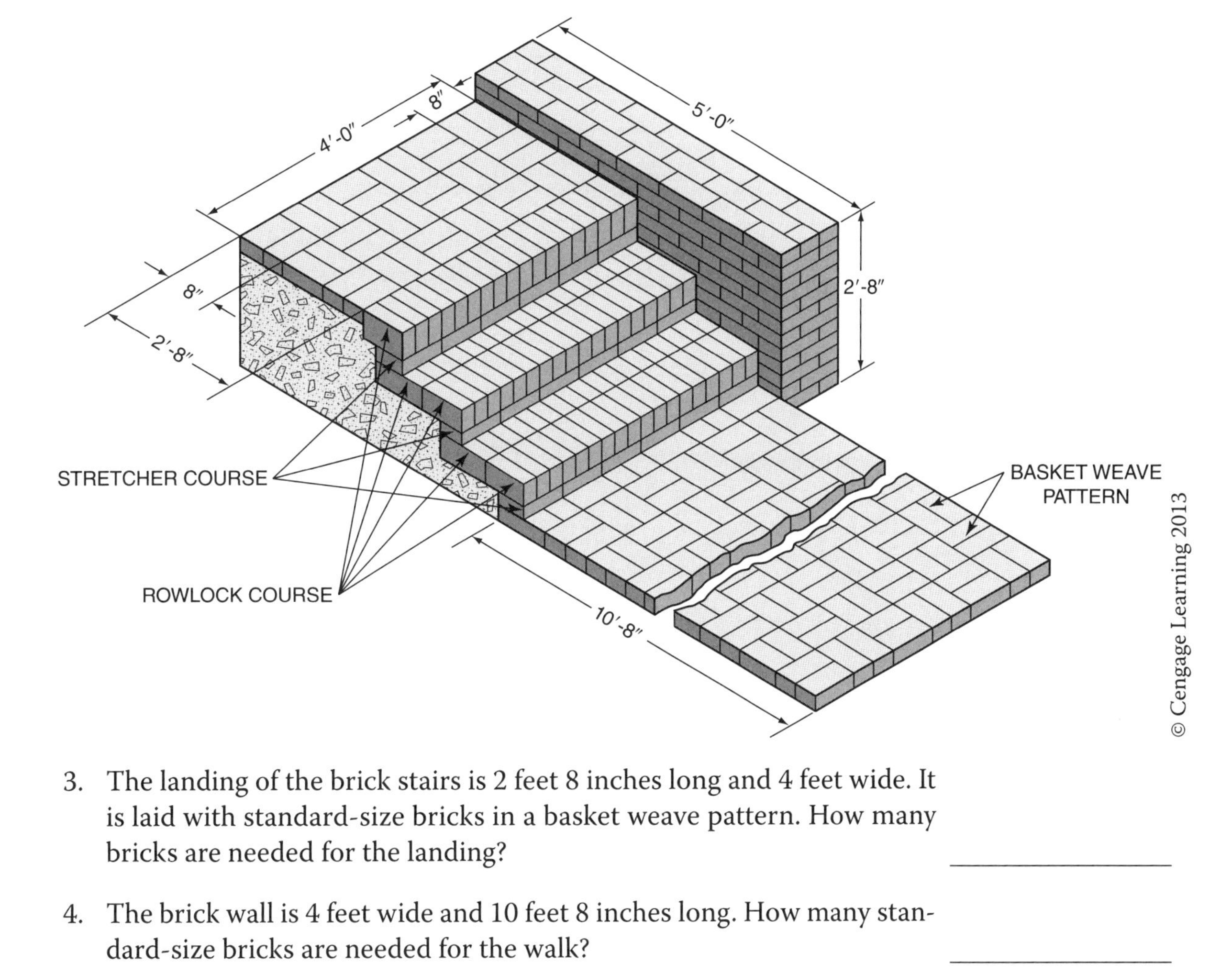

3. The landing of the brick stairs is 2 feet 8 inches long and 4 feet wide. It is laid with standard-size bricks in a basket weave pattern. How many bricks are needed for the landing? __________

4. The brick wall is 4 feet wide and 10 feet 8 inches long. How many standard-size bricks are needed for the walk? __________

5. The tread and rise of the stairs is formed by using a stretcher course and two rowlock courses. How many standard-size bricks are needed for the treads and risers? __________

APPENDIX

Reference Material for Masons

Modular Coordination

With the ever-increasing cost of construction materials and equipment, mass production has been introduced to reduce that cost. Using mass production creates the need to standardize, and the *American National Standards Institute* has approved dimensional standards for all building material and equipment.

Dimensional standards mean that masonry units, doors, windows, mechanical equipment, lumber, ceiling tile, and almost everything that is used in construction are produced so that all items "fit" together with little or no cutting or altering. To accomplish this, a standard unit of measure—4 inches—is used. The 4 inch dimension is called a *module* and dimensions that are based on the module are called *modular dimensions.* The building materials are modular-size materials.

Modular coordination means that everyone involved in construction—architects, masons, carpenters, plumbers, and electricians, to name a few—may work together knowing that the different materials of each trade can be combined and coordinated with a minimum of waste.

Modular dimensions, however, cannot be used effectively or efficiently unless building plans are carefully developed. The dimensions of a building must be exact multiples of the dimensions for the standard modular products. Since the module is 4 inches, all dimensions are multiples of 4 inches and no fractions are used.

In designing a building, the architect would make the length of a room 20 feet rather than 20 feet 2 inches. The 20-foot dimension is a multiple of 4 inches (20 feet = 240 inches), whereas a dimension of 20 feet 2 inches is not (20 feet 2 inches = 242 inches). The 20-foot dimension would allow modular masonry units such as 4″ × 2 2/3″ × 8″ bricks to be laid without wasteful cutting. It would also allow studs to be placed 16 inches on center in the partition wall.

Windows and doors are another area in which careful planning is required. Doors and windows are placed so that the courses of masonry units below and above the door or window do not have to be cut. Thus, if 8 inch high masonry units are used, the window would be placed 48 inches from the bottom of the wall rather than 50 inches from the bottom. The height of the door would be planned to be 6 feet 8 inches rather than 6 feet 3 inches. Using nonmodular dimensions, such as 50 inches or 6 feet 3 inches, would also cause a waste in jambs and sashes and the wasteful use of nonmodular-size doors and windows. The end result would be additional labor, material, and—most importantly—additional cost.

Modular Masonry Units

Most masonry units—bricks, tiles, concrete blocks—are based on the 4 inch module. The modular sizing gives the nominal size of the unit. The nominal size is the actual or manufactured size plus an allowance for mortar joint thickness. This means that the masonry unit is designed so that when the mortar joint is added, the thickness, height, and length are multiples of 4 inches.

Modular Bricks

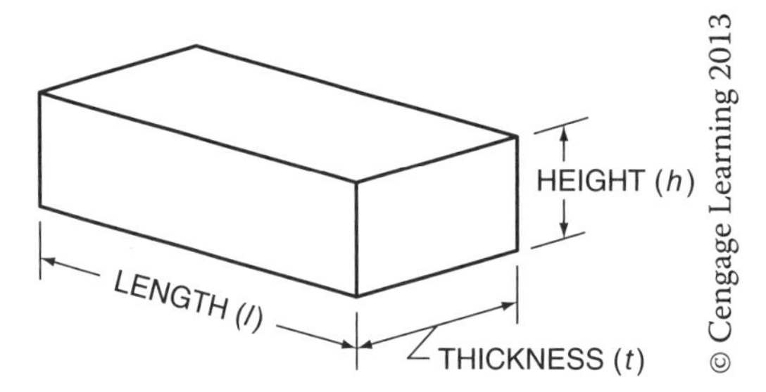

Nominal dimensions of bricks are designed so that the thickness of the brick is 4 inches, the length of the brick is a multiple of 4 inches (usually 8 inches or 12 inches), and the height is a fractional part of 4 inches (usually 1/3 of 8 inches).

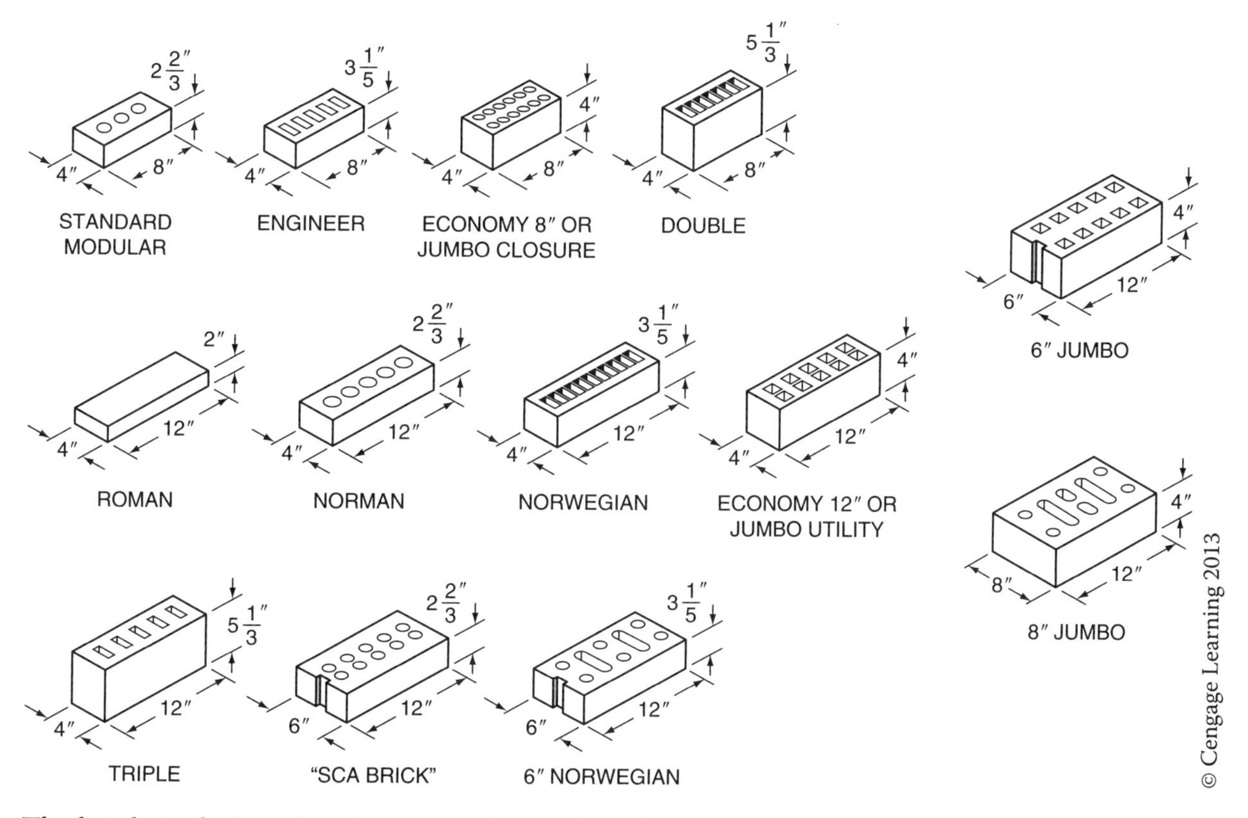

The height is designed so that a certain number of courses will bring the height of the wall to a multiple of 4 inches. For example, if the nominal height of a brick is 2 2/3 inches, three courses will bring the height to 8 inches (3 × 2 2/3″ = 8″). If the nominal height of the brick is 3 1/8 inches, five courses will bring the height to 16 inches (5 × 3 1/5″= 16″).

The following chart gives the unit designation (name of brick), the nominal dimensions, the manufactured dimensions for different joint thicknesses, and the number of courses needed to obtain a multiple of the 4 inch module.

SIZES OF MODULAR BRICKS

Unit Designation	Nominal Dimensions (inches)			Joint Thickness (inches)	Manufactured Dimensions (inches)			Modular Coursing (inches)
	T	H	L		T	H	L	
Standard Modular	4	2 2/3	8	3/8	3 5/8	2 1/4	7 5/8	3C = 8
				1/2	3 1/2	2 1/4	7 1/2	
Engineered	4	3 1/5	8	3/8	3 5/8	2 13/16	7 5/8	5C =16
				1/2	3 1/2	2 11/16	7 1/2	
Economy 8 or Jumbo Closure	4	4	8	3/8	3 5/8	3 5/8	7 5/8	1C = 4
				1/2	3 1/2	3 1/2	7 1/2	
Double	4	5 1/3	8	3/8	3 5/8	4 15/16	7 5/8	3C = 16
				1/2	3 1/2	4 13/16	7 1/2	
Roman	4	2	12	3/8	3 5/8	1 5/8	11 5/8	2C = 4
				1/2	3 1/2	2 1/4	11 1/2	
Norman	4	2 2/3	12	3/8	3 5/8	2 1/4	11 5/8	3C = 8
				1/2	3 1/2	2 1/4	11 1/2	
Norwegian	4	3 1/5	12	3/8	3 5/8	2 13/16	11 5/8	5C = 16
				1/2	3 1/2	2 11/16	11 1/2	
Economy 12 or Jumbo Utility	4	4	12	3/8	3 5/8	3 5/8	11 5/8	1C = 4
				1/2	3 1/2	3 1/2	11 1/2	
Triple	4	5 1/3	12	3/8	3 5/8	4 15/16	11 5/8	3C = 16
				1/2	3 1/2	4 13/16	11 1/2	
SCR (Structural Clay Research) Brick	6	2 2/3	12	3/8	5 5/8	2 1/4	11 5/8	3C = 8
				1/2	5 1/2	2 1/4	11 1/2	
6-Inch Norwegian	6	3 1/5	12	3/8	5 5/8	2 13/16	11 5/8	5C = 16
				1/2	5 1/2	2 11/16	11 1/2	
6-Inch Jumbo	6	4	12	3/8	5 5/8	3 5/8	11 5/8	1C = 4
				1/2	5 1/2	3 1/2	11 1/2	
8-Inch Jumbo	8	4	12	3/8	7 5/8	3 5/8	11 5/8	1C = 4
				1/2	7 1/2	3 1/2	11 1/2	

Modular Concrete Blocks

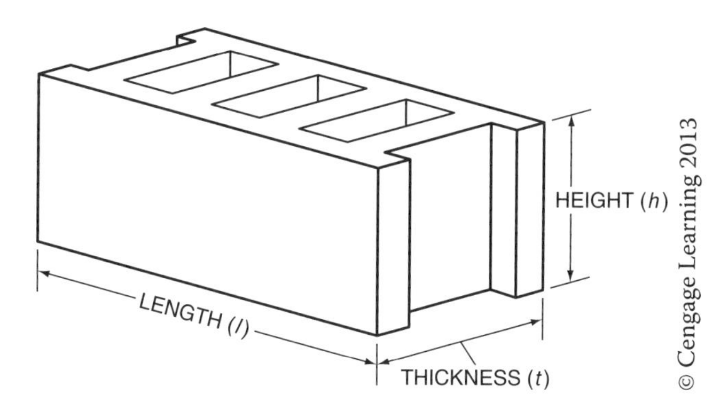

Nominal dimensions of concrete blocks are designed so that the thickness, height, and length are all multiples of 4 inches. Since the nominal size includes the mortar thickness—which is usually 3/8 inch—the actual dimensions of concrete blocks are each 3/8 inch less than the nominal dimensions.

The actual dimensions for the most popular sizes and shapes of concrete blocks are shown.

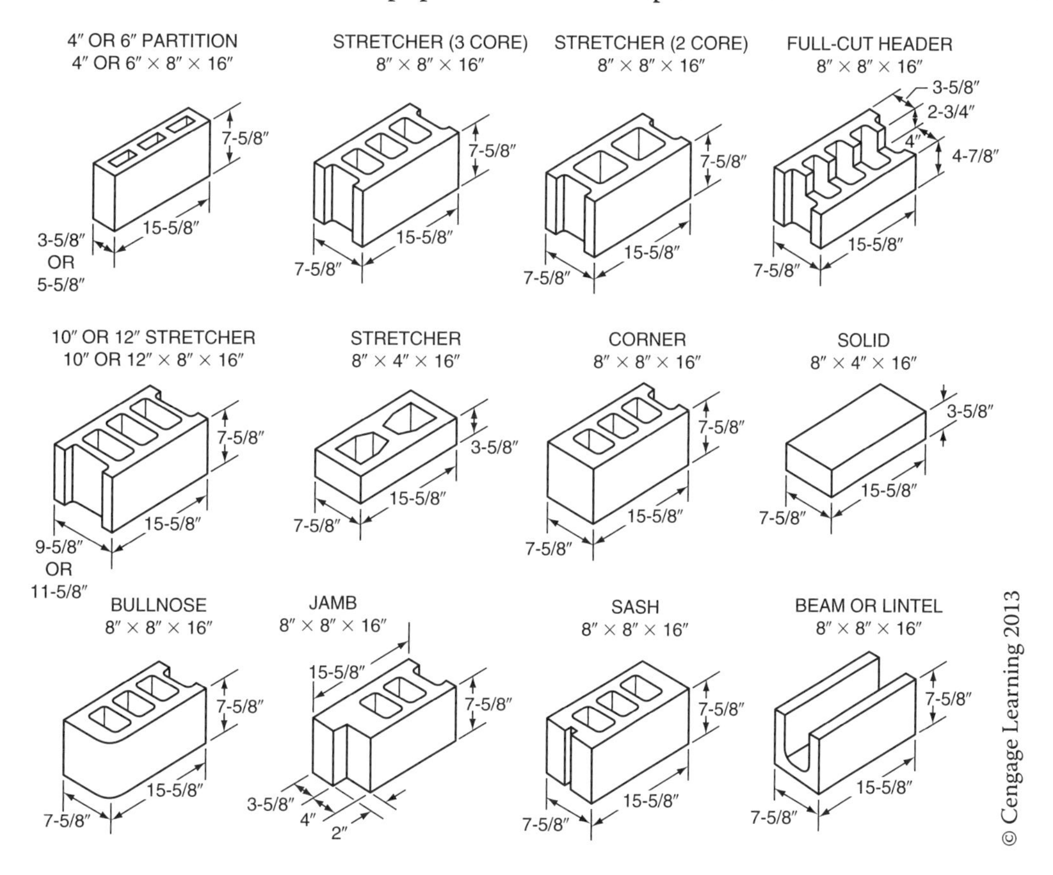

Nonmodular Masonry Units

Not all manufacturers of building materials and equipment or all contractors have accepted and are using modular coordination. For this reason, plus the fact that repairs on nonmodular buildings are still required, nonmodular masonry units are still produced and used.

Nonmodular Bricks

Nonmodular brick sizes are based on the actual or manufactured size of the brick. This means that the mason must make allowances for the mortar joint thickness. In nonmodular bricks, the actual size of the brick plus the mortar thickness is not necessarily designed to be a multiple of 4 inches.

The three most popular nonmodular bricks are the *3 inch*, the *standard* (nonmodular), and the *oversize* bricks.

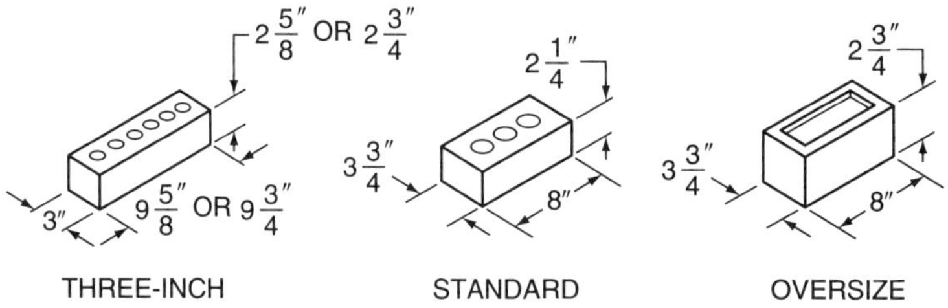

Nonmodular Concrete Blocks

The modular size of concrete blocks assumes that a 3/8 inch mortar joint is to be used. If the mortar joint thickness is different from 3/8 inch, the dimensions of the concrete block will not fit into the 4 inch module and thus can be classified as nonmodular concrete block.

Bonding

Bricks are classified according to use and position in a structure.

Single Bricks

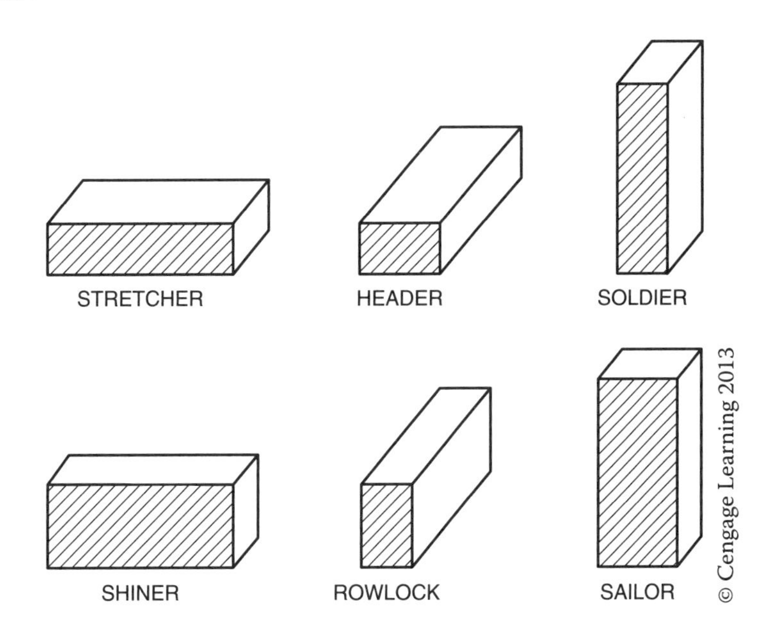

Courses of Bricks

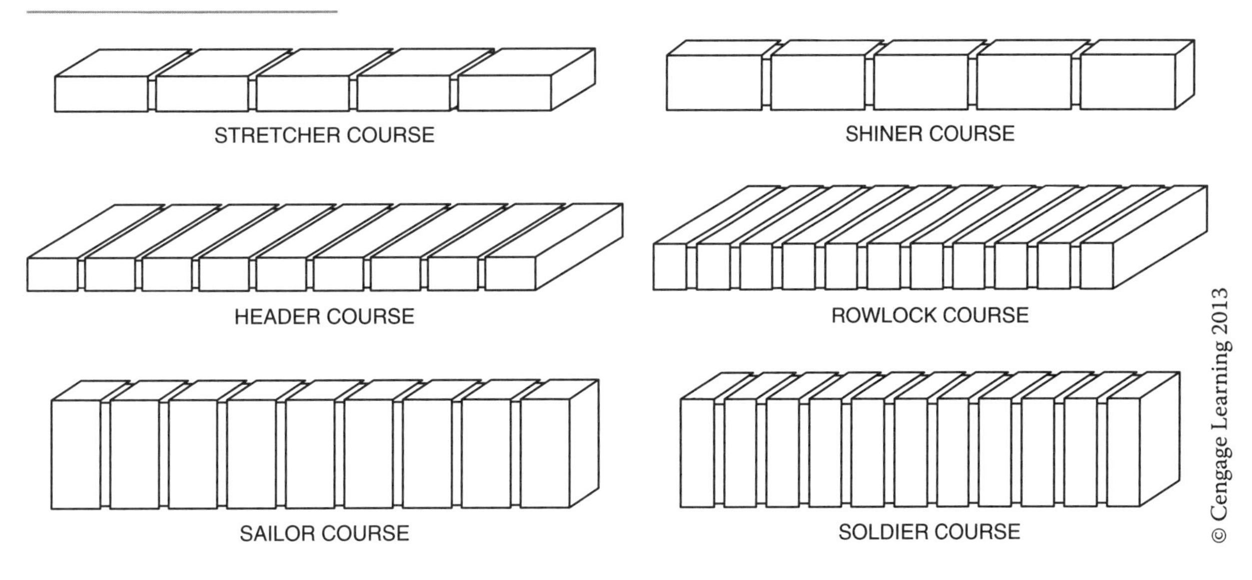

Types of Bonds

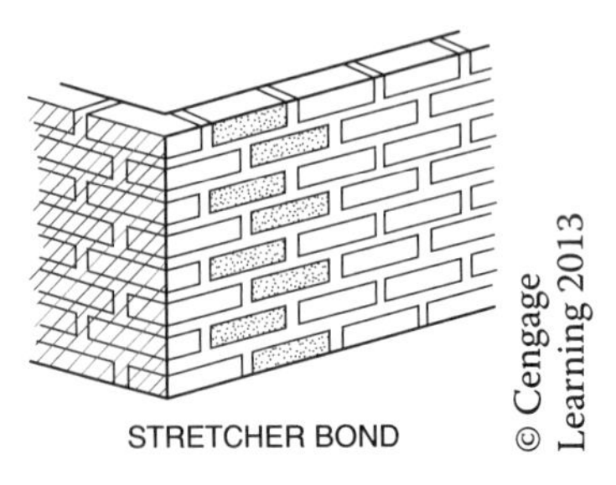

STRETCHER BOND

Also called *running bond.* All bricks are laid as stretchers with ½ lap.

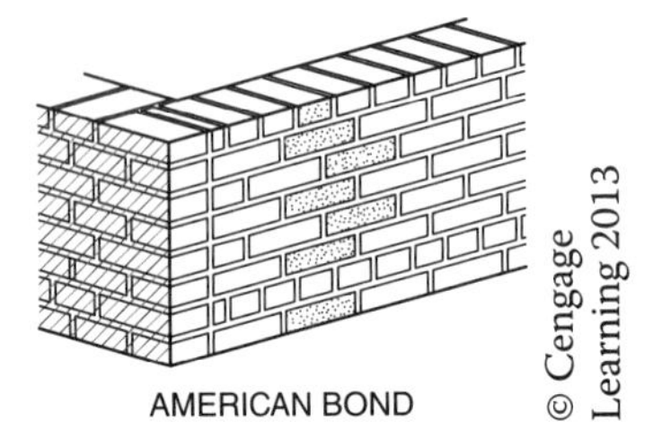

AMERICAN BOND

Also called *common bond.* Every fifth, sixth, or seventh course consists of headers. All other courses consist of stretchers.

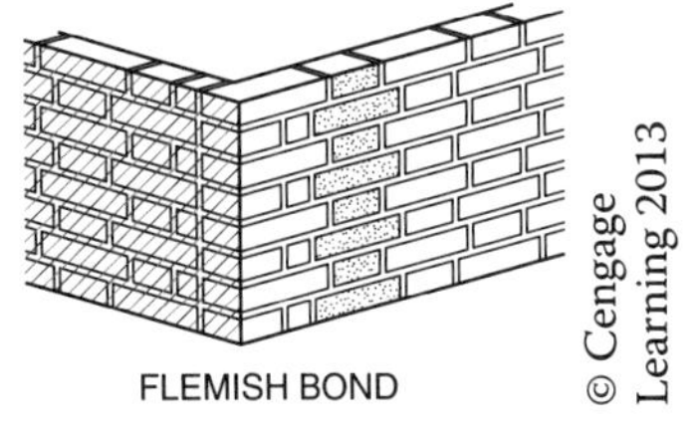

FLEMISH BOND

Each course uses alternate headers and stretcher bricks, with each header in one course centered with a stretcher in the course below and above it.

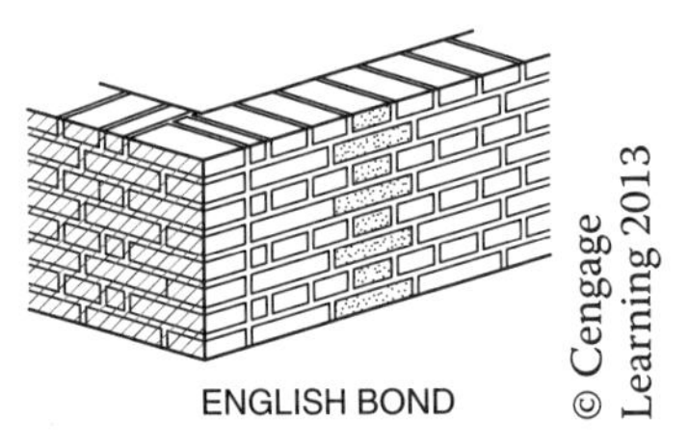

ENGLISH BOND

Alternate courses of all stretchers or all headers.

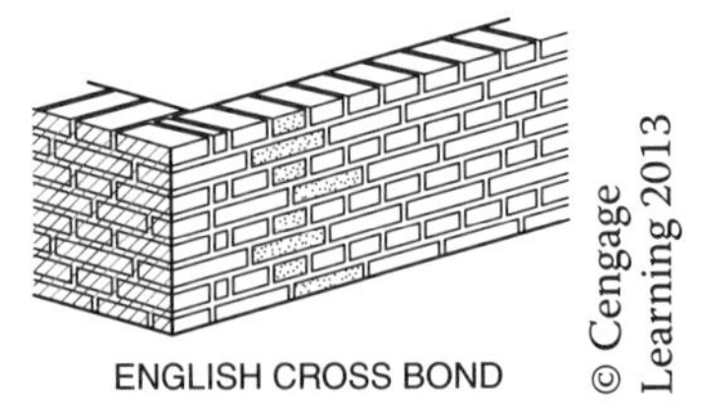

ENGLISH CROSS BOND

Also referred to as *Dutch bond* or garden wall bond. A brick pattern bond relying upon one of several combinations of stretchers and headers as well as different color combinations of the units to create noticeable diamond and diagonal line wall patterns.

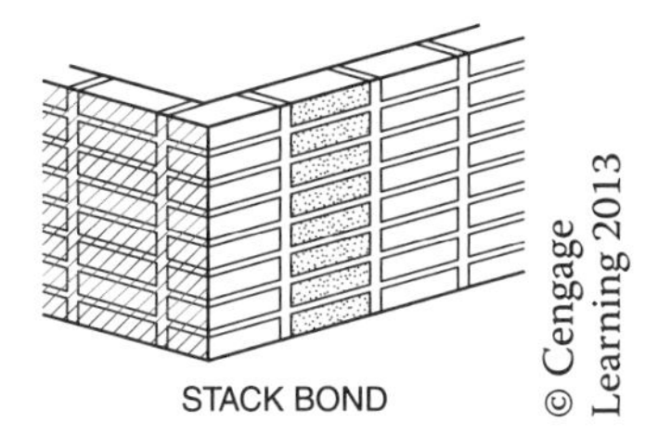

Bricks are stacked horizontally and vertically without overlapping.

Rules of Thumb

Estimating Bricks and Mortar

Bricks. Find the net wall area of square feet (length of wall multiplied by height of wall minus area of openings) and allow seven standard modular bricks (including waste) per square foot of wall area. This rule of thumb assumes that a 3/8 inch mortar joint is used.

Masonry Cement. Allow eight bags of masonry cement per 1,000 bricks. The estimate is usually rounded to the nearest larger whole bag.
Sand. Allow 3 cubic feet of sand per bag of masonry cement. Round to the nearest larger cubic yard.

Estimating Concrete Blocks and Mortar

- Determine the total linear feet by finding the outside perimeter of the building. Do not deduct for corners. This will allow for waste.
- Determine the number of concrete blocks in one course. The number of concrete blocks in one course = total linear feet × 0.75
- Determine the number of courses in the height of the wall.

  Number of courses = height of wall (in inches)/8 inches
- Determine the total number of concrete blocks in the wall.

  The total number of blocks = blocks in one course × number of courses

The allowance for specified openings in concrete block structures is determined for each opening using the steps for finding the number of concrete masonry units.

The net number of blocks required is found by subtracting the total number of blocks deducted for openings from the total number of blocks required.

The rule of thumb for masonry cement is to allow one bag of masonry cement for 30 concrete blocks. Estimates are rounded to the nearest larger whole bag.

The rule of thumb for sand is to allow 3 cubic feet of sand for each bag of masonry cement. In addition, allow 12.5 cubic feet for waste on quantities over 75 cubic feet.

ESTIMATING TABLES AND CHARTS

Mortar Type	Ratio by Volume Cement : Lime : Sand					Quantity per Cubic Foot of Mortar		
	Cement	:	Lime	:	Sand	Cement (cu. ft.)	Lime (cu. ft.)	Sand (cu. ft.)
M	1	:	¼	:	3	0.29	0.07	0.87
S	1	:	½	:	4½	0.21	0.10	0.94
N	1	:	1	:	6	0.16	0.16	0.96
O	1	:	2	:	9	0.10	0.21	0.95

MODULAR BRICKS (NOMINAL SIZE[1])

Name	Size of Bricks			Bricks per 100 Square Feet	Cubic Feet of Mortar per 1,000 Bricks	
	T	H	L		⅜-Inch Joint	½-Inch Joint
Standard Modular	4	2⅔	8	675	8.1	10.3
Engineer	4	3⅕	8	563	8.6	10.9
Closure Modular	4	4	8	450	9.2	11.7
Roman	4	2	12	600	10.8	13.7
Norman	4	2⅔	12	450	11.3	14.4
Engineer Norman	4	3⅕	12	375	11.7	14.9
Utility	4	4	12	300	12.3	15.7
6" Jumbo	6	4	12	300	19.1	24.7
8" Jumbo	8	4	12	300	25.9	33.6

[1]A nominal size of dimension is the actual dimension of the unit plus the joint thickness.

NONMODULAR BRICKS (ACTUAL SIZE)

Size of Brick (inches)			Bricks per 100 Square Feet		Cubic Feet of Mortar per 1,000 Bricks	
T	H	L	⅜-Inch Joint	½-Inch Joint	⅜-Inch Joint 8	½-Inch Joint
3	2⅝	9⅝	481	457	8.2	11.1
3	2¾	9¾	457	433	8.4	11.3
3¾	2¼	8	655	616	8.8	11.7
3¾	2¾	8	551	522	9.1	12.2

CONCRETE BLOCK AND MORTAR REQUIREMENTS FOR SINGLE-WYTHE WALLS

Nominal Size of Block	Blocks per 100 Square Feet	Cubic Feet of Mortar per 100 Square Feet
8" × 8" × 16"	112.5	6.0

SECTION I

MATHEMATICAL REFERENCE TABLES

TABLE I: POWERS AND ROOTS OF NUMBERS (1 THROUGH 100)

Number	Powers: Square	Powers: Cube	Roots: Square	Roots: Cube
1	1	1	1. 000	1. 000
2	4	8	1. 414	1. 260
3	9	27	1. 732	1. 442
4	16	64	2. 000	1. 587
5	25	125	2. 236	1. 710
6	36	216	2. 449	1. 817
7	49	343	2. 646	1. 913
8	64	512	2. 828	2. 000
9	81	729	3. 000	2. 080
10	100	1, 000	3. 162	2. 154
11	121	1, 331	3. 317	2. 224
12	144	1, 728	3. 464	2. 289
13	169	2, 197	3. 606	2. 351
14	196	2, 744	3. 742	2. 410
15	225	3, 375	3. 873	2. 466
16	256	4, 096	4. 000	2. 520
17	289	4, 913	4. 123	2. 571
18	324	5, 832	4. 243	2. 621
19	361	6, 859	4. 359	2. 668
20	400	8, 000	4. 472	2. 714
21	441	9, 261	4. 583	2. 759
22	484	10, 648	4. 690	2. 802
23	529	12, 167	4. 796	2. 844
24	576	13, 824	4. 899	2. 884
25	625	15, 625	5. 000	2. 924
26	676	17, 576	5. 099	2. 962
27	729	19, 683	5. 196	3. 000
28	784	21, 952	5. 292	3. 037
29	841	24, 389	5. 385	3. 072
30	900	27, 000	5. 477	3. 107

Number	Powers		Roots	
	Square	Cube	Square	Cube
31	961	29, 791	5. 568	3. 141
32	1, 024	32, 798	5. 657	3. 175
33	1, 089	35, 937	5. 745	3. 208
34	1, 156	39, 304	5. 831	3. 240
35	1, 225	42, 875	5. 916	3. 271
36	1, 296	46, 656	6. 000	3. 302
37	1, 369	50, 653	6. 083	3. 332
38	1, 444	54, 872	6. 164	3. 362
39	1, 521	59, 319	6. 245	3. 391
40	1, 600	64, 000	6. 325	3. 420
41	1, 681	68, 921	6. 403	3. 448
42	1, 764	74, 088	6. 481	3. 476
43	1, 849	79, 507	6. 557	3. 503
44	1, 936	85, 184	6. 633	3. 530
45	2, 025	91, 125	6. 708	3. 557
46	2, 116	97, 336	6. 782	3. 583
47	2, 209	103, 823	6. 856	3. 609
48	2, 304	110, 592	6. 928	3. 634
49	2, 401	117, 649	7. 000	3. 659
50	2, 500	125, 000	7. 071	3. 684
51	2, 601	132, 651	7. 141	3. 708
52	2, 704	140, 608	7. 211	3. 733
53	2, 809	148, 877	7. 280	3. 756
54	2, 916	157, 464	7. 348	3. 780
55	3, 025	166, 375	7. 416	3. 803
56	3, 136	175, 616	7. 483	3. 826
57	3, 249	185, 193	7. 550	3. 849
58	3, 364	195, 112	7. 616	3. 871
59	3, 481	205, 379	7. 681	3. 893
60	3, 600	216, 000	7. 746	3. 915
61	3, 721	226, 981	7. 810	3. 936
62	3, 844	238, 328	7. 874	3. 958
63	3, 969	250, 047	7. 937	3. 979
64	4, 096	262, 144	8. 000	4. 000
65	4, 225	274, 625	8. 062	4. 021
66	4, 356	287, 496	8. 124	4. 041
67	4, 489	300, 763	8. 185	4. 062
68	4, 624	314, 432	8. 246	4. 082
69	4, 761	328, 509	8. 30′	4. 102
70	4, 900	343, 000	8. 367	4. 121
71	5, 041	357, 911	8. 426	4. 141
72	5, 184	373, 248	8. 485	4. 160
73	5, 329	389, 017	8. 544	4. 179
74	5, 476	405, 224	8. 602	4. 198
75	5, 625	421, 875	8. 660	4. 217
76	5, 776	438, 976	8. 718	4. 236
77	5, 929	456, 533	8. 775	4. 254
78	6, 084	474, 552	8. 832	4. 273
79	6, 241	493, 039	8. 888	4. 291
80	6, 400	512, 000	8. 944	4. 309
81	6, 561	531, 441	9. 000	4. 327
82	6, 724	551, 368	9. 055	4. 344
83	6, 889	571, 787	9. 110	4. 362
84	7, 056	592, 704	9. 165	4. 380
85	7, 225	614, 125	9. 220	4. 397
86	7, 396	636, 056	9. 274	4. 414
87	7, 569	658, 503	9. 327	4. 481
88	7, 744	681, 472	9. 381	4. 448
89	7, 921	704, 969	9. 434	4. 465
90	8, 100	729, 000	9. 487	4. 481
91	8, 281	753, 571	9. 539	4. 498
92	8, 464	778, 688	9. 592	4. 514
93	8, 649	804, 357	9. 644	4. 531
94	8, 836	830, 584	9. 695	4. 547
95	9, 025	857, 375	9. 747	4. 563
96	9, 216	884, 736	9. 798	4. 579
97	9, 409	912, 673	9. 849	4. 595
98	9, 604	941, 192	9. 900	4. 610
99	9, 801	970, 299	9. 950	4. 626
100	10, 000	1, 000, 000	10. 000	4. 642

TABLE II

CIRCUMFERENCES AND AREAS (0.2 TO 9.8; 10 TO 99)

Diameter	Circumference	Area
0. 2	0. 628	0. 0314
0. 4	1. 26	0. 1256
0. 6	1. 88	0. 2827
0. 8	2. 51	0. 5026
1	3. 14	0. 7854
1. 2	3. 77	1. 131
1. 4	4. 39	1. 539
1. 6	5. 02	2. 011
1. 8	5. 65	2. 545
2	6. 28	3. 142
2. 2	6. 91	3. 801
2. 4	7. 53	4. 524
2. 6	8. 16	5. 309
2. 8	8. 79	6. 158
3	9. 42	7. 069
3. 2	10. 05	7. 548
3. 4	10. 68	8. 553
3. 6	11. 3	10. 18
3. 8	11. 93	11. 34
4	12. 57	12. 57
4. 2	13. 19	13. 85
4. 4	13. 82	15. 21
4. 6	14. 45	16. 62
4. 8	15. 08	18. 1
5	15. 7	19. 63
5. 2	16. 33	21. 24
5. 4	16. 96	22. 9
5. 6	17. 59	24. 63
5. 8	18. 22	26. 42
6	18. 84	28. 27
6. 2	19. 47	30. 19
6. 4	20. 1	32. 17
6. 6	20. 73	34. 21
6. 8	21. 36	36. 32
7	21. 99	38. 48

Diameter	Circumference	Area
7. 2	22. 61	40. 72
7. 4	23. 24	43. 01
7. 6	23. 87	45. 36
7. 8	24. 5	47. 78
8	25. 13	50. 27
8. 2	25. 76	52. 81
8. 4	26. 38	55. 42
8. 6	27. 01	58. 09
8. 8	27. 64	60. 82
9	28. 27	63. 62
9. 2	28. 9	66. 48
9. 4	29. 53	69. 4
9. 6	30. 15	72. 38
9. 8	30. 78	75. 43
10	31. 41	78. 54
11	34. 55	95. 03
12	37. 69	113
13	40. 84	132. 7
14	43. 98	153. 9
15	47. 12	176. 7
16	50. 26	201
17	53. 4	226. 9
18	56. 54	254. 4
19	59. 69	283. 5
20	62. 83	314. 1
21	65. 97	346. 3
22	69. 11	380. 1
23	72. 25	415. 4
24	75. 39	452. 3
25	78. 54	490. 8
26	81. 68	530. 9
27	84. 82	572. 5
28	87. 96	615. 7
29	91. 1	660. 5
30	94. 24	706. 8

Diameter	Circumference	Area
31	97. 39	754. 8
32	100. 5	804. 2
33	103. 7	855. 3
34	106. 8	907. 9
35	110	962. 1
36	113. 1	1, 017. 9
37	116. 2	1, 075. 2
38	119. 4	1, 134. 1
39	122. 5	1, 194. 6
40	125. 7	1, 256. 6
41	128. 8	1, 320. 3
42	131. 9	1, 385. 4
43	135. 1	1, 452. 2
44	138. 2	1, 520. 5
45	141. 4	1, 590. 4
46	144. 5	1, 661. 9
47	147. 7	1, 734. 9
48	150. 8	1, 809. 6
49	153. 9	1, 885. 7
50	157. 1	1, 963. 5
51	160. 2	2, 042. 8
52	163. 4	2, 123. 7
53	166. 5	2, 206. 2
54	169. 6	2, 290. 2
55	172. 8	2, 375. 8
56	175. 9	2, 463
57	179. 1	2, 551. 8
58	182. 2	2, 642. 1
59	185. 4	2, 734
60	188. 5	2, 827. 4
61	191. 6	2, 922. 5
62	194. 8	3, 019. 1
63	197. 9	3, 117. 3
64	201. 1	3, 217
65	204. 2	3, 318. 3

Diameter	Circumference	Area
66	207. 3	3, 421. 2
67	210. 5	3, 525. 7
68	213. 6	3, 631. 7
69	216. 8	3, 739. 3
70	219. 9	3, 848. 5
71	223. 1	3, 959. 2
72	226. 2	4, 071. 5
73	229. 3	4, 185. 4
74	232. 5	4, 300. 8
75	235. 6	4, 417. 9
76	238. 8	4, 536. 5
77	241. 9	4, 656. 6
78	245	4, 778. 4
79	248. 2	4, 901. 7
80	251. 3	5, 026. 6
81	254. 5	5, 153
82	257. 6	5, 281
83	260. 8	5, 410. 6
84	263. 9	5, 541. 8
85	267. 0	5, 674. 5
86	270. 2	5, 808. 8
87	273. 3	5, 944. 7
88	276. 5	6, 082. 1
89	279. 6	6, 221. 2
90	282. 7	6, 361. 7
91	285. 9	6, 503. 9
92	289. 0	6, 647. 6
93	292. 2	6, 792. 9
94	295. 2	6, 939. 8
95	298. 5	7, 088. 2
96	301. 6	7, 238. 2
97	304. 7	7, 389. 8
98	307. 9	7, 543. 0
99	311. 9	7, 697. 7

Formulas

Perimeters:

RECTANGLE

$P = 2l + 2w$

P = perimeter
l = length
w = width

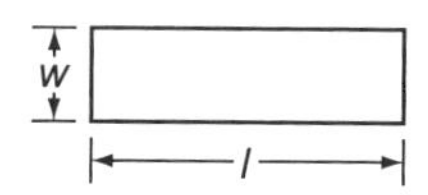

CIRCLE

$C = \pi D$

C = circumference
π = 3.1416
D = diameter

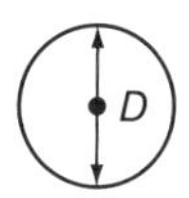

Areas:

(A = area)

CIRCLE

$A = \pi r^2$

π = 3.1416
r = radius

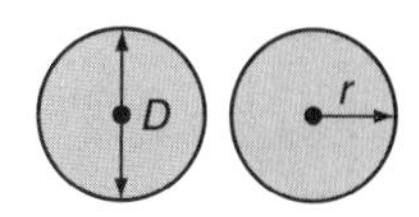

RECTANGLE

$A = lw$

l = length
w = width

SQUARE

$A = s^2$

s = length of side

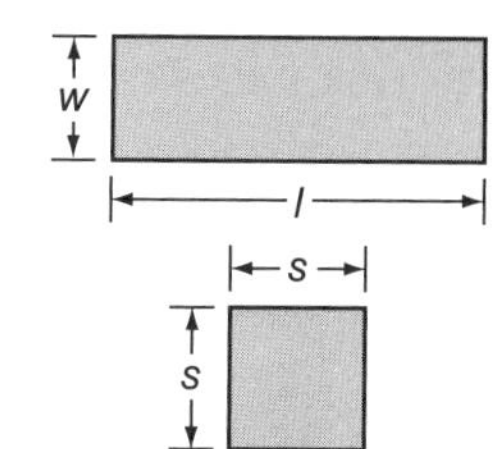

TRAPEZOID

$A = \frac{(B + b)a}{2}$

B = length of larger of the two parallel sides
b = length of smaller of the two parallel sides
a = altitude (height)

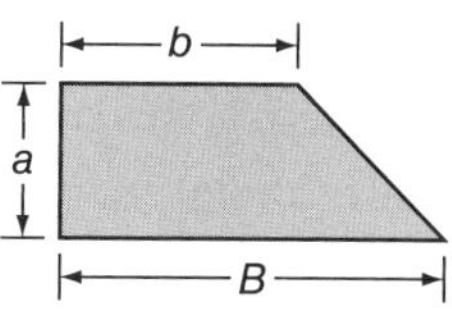

TRIANGLE

$A = \frac{bh}{2}$

a = altitude
b = base

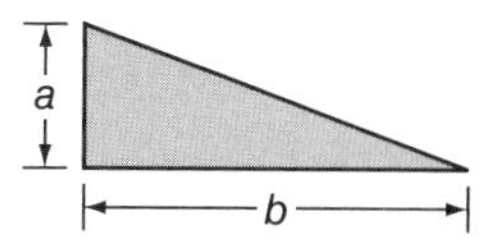

Volumes:

CYLINDER

$V = \pi r^2 l$

V = volume
π = 3.1416
r = radius
l = length

RECTANGULAR SOLID

$V = lwh$

l = length
w = width
h = height

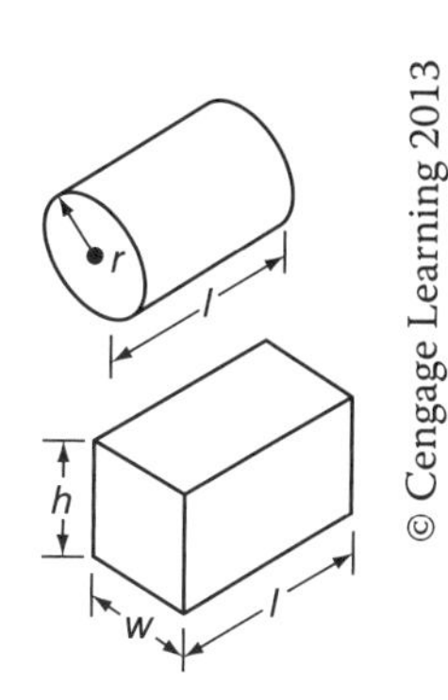

GLOSSARY

Actual size The manufactured actual dimensions of a masonry unit or building material; the dimensions of a masonry unit excluding mortar joint thickness.

Aggregate A reference to crushed stone of various sizes and masonry sand: ingredient of mortar, grout, and concrete.

American bond Also called common bond. A brick pattern bond whereby every fifth, sixth, or seventh course consists of bricks laid in the header position and the courses between are laid in the stretcher position.

Angle iron A section of structural iron formed in the shape of the letter "L" whose two legs form a right angle. Used as a steel lintel to support masonry units above masonry wall openings.

Artificial stone Also called manufactured stone. A manufactured product containing Portland cement and color pigments that is formed by casting in molds to resemble a variety of types of natural stone. It can be applied as a veneer, bonded to almost any structurally sound surface.

Ash dump door A steel or cast iron fireplace accessory installed to have a flush fit on the floor of the fireplace inner hearth. It allows discarding ashes into an ash pit below the firebox.

Back hearth Also called the inner hearth. The flat, bottom part of the firebox located behind the fireplace face opening. The part of a firebox supporting firewood and a grate. It is usually built with special firebricks.

Backfill Coarse dirt, crushed stone, or other approved material placed on the outside of basement or foundation walls below grade.

Basket weave bond A brick pattern bond in which bricks are paired perpendicular to adjoining pairs.

Bat Another term for a half-length brick.

Bearing wall A masonry wall that supports loads other than its own weight, such as the structural weights of floors, walls, and roofs of a building and the loads imposed on them such as occupants, furnishings, and equipment.

Bed joints The mortar joints below bricks, blocks, and stones on which they are bedded or supported.

Block Also called cinder block, cement block, concrete masonry unit or CMU. Masonry building units made from Portland cement and small size aggregates mixed with water and formed in a mold.

Brick A masonry building material used for the construction of walls and pavement surfaces. Pulverized clay or shale mixed with water and formed into a desired shape before drying and firing in a kiln to produce a durable ceramic product.

Brick veneer Exposed brickwork supporting no weight other than its own and either anchored to or adhered to a back-up wall constructed of other material, typically being wood framework, steel framework, or CMUs.

Building brick Also called common brick. A brick masonry unit that is made primarily for use where appearance is not considered. It is not specially treated for texture or color.

Bullnose block A CMU having a rounded corner at one or two ends that is used at the ends of walls to provide safety, resist chipping, and enhance appearance.

Cavity wall A masonry wall consisting of an inner and outer wythe separated by an air space, possibly containing thermal insulation, not less than 2″ or more than 4½″ joined with corrosion-resistant horizontal joint reinforcement.

Cement A substance for binding other materials together; any substance that causes materials to adhere to one another, such as Portland cement, masonry cement, stucco, mortar, or plaster of Paris.

Cement mortar Also called mortar cement. A mixture of Portland cement or blended hydraulic cement and plasticizing materials having a lower air content than masonry cements. When mixed with sand and water, it bonds masonry units together.

Ceramic mosaic Small pieces of ceramic tiles grouped to create artistic images and patterns.

Chase A recessed or enclosed area of a masonry wall intended to provide space for electrical panels and conduit, heating and cooling ductwork, and plumbing.

Cheek wall Wall on each side of a set of steps that is higher than the platform and can possibly serve as guard railing.

Chimney A structure venting the combustion by-products of fireplaces, wood stoves, and oil- or gas-fired appliances to the outside air while protecting the building and its occupants from fire and the life-threatening effects of carbon monoxide poisoning.

Column A masonry structure whose width does not exceed four times its thickness and whose height exceeds four times its least lateral dimension.

Common bond Also called American bond. A brick pattern bond whereby every fifth, sixth, or seventh course consists of bricks laid in the header position and the courses between are laid in the stretcher position.

Common brick Also called building brick. A brick masonry unit that is made primarily for use where appearance is not considered. It is not specially treated for texture or color.

Composite wall A masonry wall consisting of two wythes of masonry units having different strength characteristics, connected with corrosion-resistant horizontal joint reinforcement or brick headers and acting as a single wall in resisting forces and lateral loads.

Concrete A proportioned mixture of cement, sand, gravel, and water, which hardens by a chemical reaction called hydration, a reaction between water molecules with the mineral ingredients.

Concrete block Also called cement block, cinder block, concrete masonry unit, or CMU. A precast hollow or solid unit containing Portland cement and finely crushed limestone mixed with water that is used in the construction of foundation walls and masonry buildings.

Concrete masonry unit Also called CMU. A general term that refers to either concrete blocks or lightweight blocks.

Control joint A vertical joint controlling the location of separation in a concrete masonry wall resulting in the dimensional changes of the building materials.

Corbel A projection of the face(s) of brick(s) or other masonry units beyond the face of the wall below, forming a stepped-out appearance.

Corner block Also called jamb block. A CMU having at least one flat end used at the exposed ends of masonry wall corners, jambs, and wall openings.

Course A single row of bricks or other masonry units laid in a wall; to arrange in a row.

Double brick A modular-size brick with nominal dimensions of $4'' \times 5\frac{1}{3}'' \times 8''$.

Double corner block Also called a pier block. A CMU designed for use in laying piers, pilasters, or for any other purpose where both ends of the block are visible. In this type of block, both corners are flat and finished.

Double wythe wall A masonry wall whose bed depth consists of two units of masonry, two walls joined back-to-back with horizontal joint reinforcement and/or grout and rebar to form a single wall.

Downdraft The vertical movement of air down a chimney flue. An expression describing the inability for combustion by-products to escape up a flue and to the outside air, caused by poor design or obstructions and having the potential to cause smoke-filled rooms and carbon monoxide poisoning.

Draft The movement of air created by a variation in air pressure resulting from a mass of warm air being less dense than cold air. In a chimney flue it is the result of the differences in the density of hot air in the flue and the cooler atmospheric air.

Dry well A deep hole filled with crushed stone that allows drainage water to accumulate until it is absorbed by ground soil.

Dutch bond Also referred to as English cross bond or garden wall bond. A brick pattern bond relying upon one of several combinations of stretchers and headers as well as different color combinations of the units to create noticeable diamond and diagonal line wall patterns.

Economy 8 brick A modular-size brick with nominal dimensions of 4″ × 4″ × 8″. (Also referred to as Jumbo closure bricks.) Modular construction permits a height of 16″ with four courses of economy brick bed in approximately ⅜″ bed joints.

Economy 12 brick A modular-size brick with nominal dimensions of 4″ × 4″ × 12″. (Also referred to as Jumbo utility brick.)

Efflorescence A chalky white deposit on the exposed surface of masonry walls caused by moisture or water dissolving soluble salts present in building materials.

Elevation view A two-dimensional graphic scale representation of a single side of a building or structure.

Engineer brick Also called engineered brick or oversized brick. A modular-size brick with nominal dimensions of 4″ × 3⅕″ × 8″.

English bond A brick pattern bond consisting of alternating courses of headers and stretchers having each unit aligned centered with the unit below and above it.

English cross bond Also referred to as Dutch bond or garden wall bond. A brick pattern bond relying upon one of several combinations of stretchers and headers as well as different color combinations of the units to create noticeable diamond and diagonal line wall patterns.

Estimating The process of judging or calculating approximate construction costs, including but not limited to the amount of material, labor, and equipment required for a given job.

Expansion joint A horizontal or vertical separation of a brick wythe that is filled with an elastic sealant and backer rod permitting the expansion of brick walls caused by thermal movements or increasing volume of bricks.

Face Also called face side. The front or exposed surface of a wall.

Facing brick A term describing one of several grades of bricks that is specially treated in the molding or firing processes to produce face and end surfaces having desired textures and colors. Used for exterior and interior walls where appearance is an architectural requirement.

Faced wall A masonry wall in which a different material such as ceramic tile, thin brick veneer or precast stone is adhered to one or both sides of the wall.

Firebrick A brick, usually made of fireclay or other highly siliceous material, that is especially made to withstand the effects of high heat without failure. Used for the exposed surfaces of a fireplace firebox.

Fireclay Also called refractory clay. Clay mined at great depths, capable of withstanding high temperatures used to make firebrick and flue liners.

Flashing A water-impermeable material such as aluminum, copper, stainless steel, PVC, and rubber used as a water collection membrane in a wall system. It diverts both permeating water and water condensation within the wall to the exterior of the wall, protecting building materials from rot, corrosion, and performance failure.

Flemish bond A brick pattern bond formed by alternating headers and stretchers in each course having each unit aligned centered with the unit below and above it.

Flue An enclosed passageway, such as a masonry chimney or a chimney pipe, intended to contain combustion by-products and direct them safely to the outside air.

Flue lining A rectangular-shaped or round-shaped, hollow chimney lining made of fireclay and intended to contain the combustion by-products and protect the masonry chimney walls from heat. Typically available in lengths of two feet.

Footing Also called a footer. The structural support at the base of a wall, typically consisting of concrete, that uniformly distributes the weight of the building or structure over a wider area to prevent settling.

Footing forms Wooden, PVC, or steel assemblies placed to form a footing to the desired shape and size until the concrete hardens.

Fore hearth Also called the outer hearth. The part of the fireplace hearth that extends beyond the inner hearth, projecting out from the fireplace face for the purpose of protecting the living area floor from hot embers.

Form Also called formwork. A mold consisting of temporarily braced framing lumber, reusable forming systems, or permanent forms used to support and contain concrete to assume a desired shape until it hardens.

Foundation that part of a building or structure below the first floor framing supporting the building or structure.

Foundation wall A wall below the floor nearest grade level, partially or mostly below grade level, that serves as a support for a wall or other structural part of the building above grade level.

Grade Also called finish grade or grade level. A term referring to the ground level around a building.

Grout A mixture of Portland cement and aggregates to which sufficient water is added to produce a pouring consistency without separation of the ingredients, used as fill between masonry wall wythes and the open cells of CMUs and other masonry units to strengthen masonry walls.

Header brick The position of a brick observed in a wall having its bottom side bedded in mortar and its end oriented with the face of the wall.

Header block A concrete block that has a formed shelf on one of its two sides to facilitate bonding with brick laid in the header position or for additional support at the perimeter of a concrete slab.

Header course A course of bricks laid in the header position, typically tying together the two wythes forming 8″ brick walls or tying a 4″ brick wall into a CMU back-up wall.

Header bond A brick pattern bond formed by orienting brick at the face of the wall in the header position.

Head joint The vertical mortar joint between the ends of two adjacent bricks or CMUs.

Hearth The combined area including the outer hearth and inner hearth of a fireplace.

Hydrated lime The material resulting from the chemical reaction of quicklime and water. An ingredient of many masonry mortars.

I beam A steel beam whose cross-sectional area resembles the letter "I," used as a component of steel framework or as a masonry lintel.

Inner hearth Also called the back hearth. The flat-bottom part of the firebox located behind the fireplace face opening. The part of a firebox supporting firewood and a grate. It is usually built with special firebricks.

Isolated footing A footing separate from the main support footing of a building. It is used to support such structures as a masonry pier, column, and chimney.

Jamb The exposed end of a wall opening.

Jamb block A CMU having a square, flat end that is used at wall openings.

Joint Also called mortar joint. The space between adjacent bricks, CMUs, stones, or other masonry units filled with mortar and creating a mortar bond between individual masonry units.

Jumbo closure brick A modular-size brick with nominal dimensions of 4″ × 4″ × 8″. (Also referred to as Economy 8 brick.)

Jumbo utility brick A modular-size brick with nominal dimensions of 4″ × 4″ × 12″. (Also referred to as economy 12 brick.)

L-block Also called offset corner block. A specially shaped CMU that is used at the ends of a CMU corner. Used as the end block on each course, its offset end width creates corners having a half-lap, running bond pattern.

Lally column A cylindrical-shaped steel member that is used as a support for girders or other beams. It is sometimes filled with concrete.

Landing A platform used in step construction to change the direction of the steps or to break the run.

Lap The distance that one masonry unit or other material extends or projects onto another.

Lath An open-mesh metal screening secured to a back-up wall over which stucco is applied.

Lightweight block A CMU containing Portland cement and lightweight aggregates such as expanded shale, resulting in a lighter weight block than concrete block and having improved properties when compared to a concrete block.

Lime A highly infusible, caustic, whitish substance that is produced by the reaction of heat on limestone, shells, or other forms of calcium carbonate.

Linear dimension A single measurement referring to the length, width, or height of an object.

Lintel A concrete masonry product, structural or engineered wood component, stone, or steel I-beam or angle iron placed horizontally across a wall opening to support loads above the opening. Concrete masonry lintels are steel reinforced, precast masonry units used to support loads above wall openings and to bridge adjacent footings. They are available in standard block widths/heights and several lengths.

Manufactured size The actual dimensions of the masonry unit; the dimensions without accounting for the width of mortar joint.

Mason A craftsperson skilled in laying bricks and CMUs (brickmason) or one skilled in laying stones (stonemason).

Masonry A general term for anything constructed of bricks, CMUs, structural tiles, concrete or similar materials; the work performed by a person who works with bricks, CMUs, structural tiles, or concrete.

Masonry cement A mixture of Portland cement or blended hydraulic cement and plasticizing materials whose air content varies widely between manufacturers, recognized for their good workability when sand and water are added to make mortar.

Modular brick A brick in whose dimensions when including mortar joints are nominal, based on a 4″ module.

Modular dimensional standards The dimensional standards that are based upon a standard unit of measure of 4″, known as the module. These standards are industry-approved standards facilitating modular construction intended to ease the assembly of various components.

Modular masonry Masonry construction in which the size of the building material, such as bricks, tiles, or concrete blocks, is based on a module and dimensions are multiples of 4 inches. It is intended to ease the assembly of different masonry materials when building composite or cavity walls.

Module The unit of measure, 4 inches, that is the industry standard for all modular building materials and equipment.

Mortar A mixture of cementitious materials, sand, and water. It is used as to bond brick, block, and stone to create masonry walls.

Mortar bed A layer of mortar in which masonry units are seated or bedded. The purpose is to provide a contour-formed base allowing the formation of a mortar bond between the bedded masonry unit and the masonry below it.

Mortar bond A measure of the resistance to separation of mortar and masonry units.

Mortar joint The mortar-filled space between individual masonry units forming a structural bond between them.

Mosaic A combination of small, colored stones; glass; or other material arranged to form a decorative pattern. The pieces are inlaid, usually in cement or stucco.

Nominal size An expression of an object's approximate size rather than its actual size used for identification purposes. For masonry units it is typically the actual dimensions of the unit plus the face width of one ⅜″-wide bed joint and one ⅜″-wide head joint.

Non-modular bricks Bricks whose dimensions are not necessarily compatible with the modular masonry system. Their dimensions are typically slightly greater than the dimensions of comparable modular masonry bricks.

Norman brick A modular-size brick with nominal dimensions of 4″ × 2⅔″ × 12″.

Norwegian brick A modular-size brick with nominal dimensions of 4″ × 3⅕″ × 12″.

Oversize brick Also called oversized brick, engineer size brick or engineered size brick. A brick whose approximate dimensions are 7⅝″ long, 3½″ wide, and a face height of 2 13/16″. Modular construction permits a height of 16″ with five courses of oversize modular brick bed in approximately ⅜″ bed joints.

Partition wall Also called screen wall. A non-load bearing wall separating two room areas, typically constructed of 6″ CMUs or 8″ CMUs.

Pattern bond The visual pattern created at the face of a masonry wall by the arrangement of masonry units in various brick positions.

Paving brick Brick having properties capable of withstanding pedestrian traffic and/or vehicular traffic; used on walkways, driveways, and roadways.

Pier Unattached vertical column of masonry work that is not bonded to a masonry wall. It serves as a supporting section for arches and lintels.

Pilaster A rectangular column attached to a wall or pier and projecting from it.

Plan view Also called working drawing. A graphic or pictorial representation of elements of a project showing their design, location, and dimensions.

Plasticizing materials Materials such as ground limestone, hydrated or hydraulic lime; an ingredient in masonry cements, mortar cements, and Portland Cement-Lime mixtures intended to enhance the workability of mortar.

Platform The horizontal surface located at the top of a stairway. It is usually constructed before the steps are built. Also called a stoop.

Porch A sheltered area at the entrance of a building.

Portland cement A hydraulic cement consisting of silica, lime, and alumina mixed in proper proportions and then burned in a kiln. An ingredient of masonry mortars, grout, and concrete.

Quarry tile A machine-made, unglazed floor tile. It is made from natural clays and is generally a very dense, dark-reddish tile.

Queen-size brick A brick having a length of 7⅝″, face height of 2¾″, and bed depth of 2¾″, approximately ¾″ narrower than most other bricks.

Reinforced concrete Concrete that has been strengthened by iron or steel bars embedded in it.

Reinforced masonry wall A masonry wall strengthened with steel reinforcement rods and grout.

Retaining wall Any wall erected to hold back or support a bank of earth.

Rise The vertical distance through which anything rises, such as the rise of a step or the rise of a roof.

Roman brick A modular-size brick with nominal dimensions of 4″ × 2″ × 12″. Modular construction permits a height of 16″ with eight courses of Roman brick bed in approximately ⅜″ bed joints.

Rowlock brick The position for a brick observed in a wall having its end oriented with the face of the wall and the backside of the brick bedded in mortar.

Rowlock course A horizontal row of bricks laid in the rowlock position, having the face side of the brick oriented to be the finished top of the wall. A rowlock course is often used as the exposed top of brick walls, columns, and piers.

Rubble stone A stone as it comes from the quarry, field, or river bed. A stone not dressed or cut. Irregular-shaped stones laid in a random pattern create walls referred to as rubble stone walls.

Running bond Also called stretcher bond. A widely used pattern bond for both brick walls and CMU walls whereby each brick or CMU is laid in the stretcher position. The bricks or CMUs of alternate courses form a uniform overlap and plumb-aligned head joints on alternate courses.

Sailor brick The position for a brick observed in a wall having its top or bottom side oriented with the face of the wall and its end bedded in mortar. Primarily used with solid brick to create decorative brick patterns.

Scaffold An elevated, and usually temporary, platform used for supporting the safe limited combined weight of craftspersons, tools, equipment, and building materials while working on a building.

SCR brick (Structural Clay Research brick) A modular-size brick with nominal dimensions of 6″ × 2⅔″ × 12″. This brick was developed to provide increased efficiency, better workmanship, and more production in general masonry construction.

Self-supporting steps Steps supported by footings separate from the main wall footing.

Single wythe A term describing a masonry wall whose mortar bed depth is equivalent to one masonry unit.

Soldier brick The position for a brick having its end bedded in mortar and its face side oriented with the face of the wall.

Span The horizontal distance between abutments; the width of a wall opening.

Stack bond A pattern bond for brick and block masonry whereby all units are aligned directly above lower units, forming continuous vertical head joint alignments between adjacent units of each course.

Standard concrete masonry unit CMUs having a standard length of 15⅝″, a standard height of 7⅝″, and one of six standard widths, those being 3⅝″, 5⅝″, 7⅝″, 9⅝″, 11⅝″, and 13⅝″. Block used to create a running bond pattern.

Standard modular brick The most frequently used type of modular brick; brick with nominal dimensions of 4″ × 2⅔″ × 8″. Modular construction permits a height of 16″ with six courses of standard size modular brick bed in approximately ⅜″ bed joints.

Standard non-modular brick One of the most popular types of non-modular bricks; brick with actual dimensions of 3¾″ × 2¼″ × 8″.

Step One unit of a stairway composed of one tread and one riser. It may exist alone or in a series or set to form a stairway.

Story pole (story rod) A wooden pole or rod, typically a 1″ × 2″ wood furring strip, that has been marked to show masonry construction details for a specific wall, markings can include masonry coursings, bottoms and tops of wall openings, and other wall detail markings.

Stretcher block Also called a standard CMU. CMUs having a standard length of 15⅝″, a standard height of 7⅝″, and one of six standard widths, those being 3⅝″, 5⅝″, 7⅝″, 9⅝″, 11⅝″, and 13⅝″. Block used to create a running bond pattern.

Stretcher brick The position for a brick observed in a wall having its bottom side bedded in mortar and its face side oriented with the face of the wall.

Stretcher bond Also called running bond. A widely used pattern bond for both brick walls and CMU walls whereby each brick or CMU is laid in the stretcher position while the bricks or CMUs of alternate courses form a uniform overlap and plumb-aligned bond.

Stretcher course A horizontal row of bricks having each brick laid in the stretcher position.

Terra-cotta A clay product that is used for ornamental work on the exterior of buildings, drain tile, and some other water resistant/heat resistant building materials. It is made of hard-baked clay in variable colors and has a fine glazed surface.

Three-core block A CMU having three hollow cores or cells separated by cross webs.

Three-inch brick One of the most popular types of non-modular bricks; brick with actual dimensions of 3″ × 9⅝″ or 9¾″ × 2⅝″ or 2¾″.

Tilt-up panel A precast section of concrete, cast in forms at the jobsite, removed from the forms after curing, and raised or tilted by a crane into its final position.

Traprock A graded or sized, uncolored, natural rock that is used as an underlay between two layers of asphalt.

Tread The horizontal surface of a step supporting pedestrian foot traffic.

Triple brick A modular-size brick with nominal dimensions of 4″ × 5⅓″ × 12″.

Triple-wythe wall A masonry wall whose bed depth consists of three units of masonry, three walls joined back-to-back with horizontal joint reinforcement or grout and rebar to form a single wall.

Two-core block A CMU having two hollow cores or cells separated by cross webs.

Veneer wall A wall with a masonry facing that is anchored to the backing but is not bonded to act as a load-bearing wall. Also called brick veneer wall or anchored brick veneer wall.

Wall footing That part of the foundation consisting of concrete designed to support the foundation walls and transmitting the structure's weight to the soil below.

Wash gravel A coarse aggregate, either gravel or stone, ranging in size from ¼″ up to 1½″ that has been washed with water to make sure that it is free of earth or organic matter. Used as a filler under concrete floors and slabs.

Wythe A term expressing the bed depth of a masonry wall or thickness of a masonry wall in masonry units: each unit of bed depth is considered a wythe; a row or thickness of masonry units in a wall.

Working drawing Also called plan view. A graphic or pictorial representation of elements of a project showing their design, location, and dimensions.

Workpiece In masonry, the building or structure that is being built, restored, or repaired with masonry units.

ANSWERS
TO ODD-NUMBERED PROBLEMS

Section 1
Whole Numbers

Unit 1
Addition of Whole Numbers

1. 148 feet
3. $1,450
5. 600 blocks
7. 40,300 bricks

Unit 2
Subtraction of Whole Numbers

1. $17
3. 12 hours
5. 12 feet

Unit 3
Multiplication of Whole Numbers

1. $720
3. 680 bricks
5. 58 blocks
7. 35 hours

Unit 4
Division of Whole Numbers

1. 125 bricks
3. 510 standard-size bricks
5. 69 square feet per hour
7. 109 square feet

Unit 5
Combined Operations with Whole Numbers

1. 7 weeks
3. 300 square feet
5. 3,306 bricks

Section 2
Common Fractions

Unit 6
Addition of Common Fractions

1. a. $\frac{1}{8}$ inches
1. c. $\frac{1}{4}$ inches
3. $34\frac{5}{6}$ hours
5. 8 inches

Unit 7
Subtraction of Common Fractions

1. a. $7\frac{5}{8}$ inches
3. 1 foot $10\frac{15}{16}$ inches
5. $4\frac{1}{2}$ hours

Unit 8
Multiplication of Common Fractions

1. a. $2\frac{1}{4}$ inches
1. c. $\frac{1}{15}$ yard
3. 10 feet $3\frac{1}{4}$ inches
5. 2 feet 6 inches

Unit 9
Division of Common Fractions

1. a. $\frac{1}{4}$ inch
1. c. $\frac{1}{16}$ inch
3. 4 feet $1\frac{1}{2}$ inches

Unit 10
Combined Operations with Common Fractions

1. a. $\frac{1}{2}$ inch maximum
   c. $\frac{1}{4}$ inch minimum
3. 32 $\frac{3}{8}$ inches
5. 12 inches
7. $47\frac{5}{8}$ inches
9. $9\frac{1}{16}$ inches
11. $\frac{1}{16}$ inch
13. $\frac{1}{8}$ inch
15. $\frac{1}{4}$ inch
17. $\frac{3}{8}$ inch
19. $19\frac{3}{8}$ inches
21. 29 stretcher courses below the rowlock cap
23. 6 feet $2\frac{1}{4}$ inches
25. $787\frac{1}{2}$ pounds

Section 3
Decimal Fractions

Unit 11
Addition of Decimal Fractions

1. 52.50
3. 1.08 inches
5. 22.25 hours
7. $7,481.75

Unit 12
Subtraction of Decimal Fractions

1. 113.76
3. 9.7 yards
5. $473.25
7. 24 inches

Unit 13
Multiplication of Decimal Fractions

1. 5.625 inches
3. 61.6
5. 58.5 inches
7. 19.50 inches

Unit 14
Division of Decimal Fractions

1. 2.1 inches
3. $28.07
5. 75.04
7. $253.70
9. 2.75 inches

Unit 15
Decimal Fractions and Common Fraction Equivalents

1. 0.625
3. 0.3125
5. $\frac{5}{16}$
7. 2.25 inches
9. $2\frac{3}{4}$ pounds

Unit 16
Combined Operations with Decimal Fractions

1. 10.5
3. 8.28
5. 83.75 pounds
7. $1,218.85

Section 4 Percentages, Averages, and Proportions

Unit 17
Calculating Percentages

1. 50
3. 9
5. 39.6
7. 900 bricks
9. 60%
11. 12%
13. 51%

Unit 18
Interest

1. $7,670
3. $432
5. $2.799.25

Unit 19
Averages

1. 12,215 bricks
3. 61 miles
5. $1,940.25

Unit 20
Calculating Ratios and Proportions

1. 10:2
3. 8:18
5. 1⁄125
7. 22,000/176
9. 15,000 engineered-size bricks
11. 5 feet 10 inches
13. 80 hours
15. 20 feet
17. 9 square feet
19. 19 rebars

Section 5 Powers and Roots

Unit 21
Powers

1. 4
3. 16
5. 49
7. 125
9. 27

Unit 22
Roots

1. 1.414
3. 5
5. 12
7. 4.796
9. 3

Unit 23
Combined Operations with Powers and Roots

1. 6
3. 10 inches
5. 10.55 feet
7. 47.99 feet = 47 feet 11⅞ inches

Section 6 Measure

Unit 24
Metric Measure

1. 100 centimeters
3. 260 millimeters
5. 6.25 meters

Unit 25

Rule or Tape Having 1/16-Inch Graduations

1. 8 parts
3. a. 6 1/8 inches
3. b. 6 1/4 inches
3. c. 6 3/8 inches
3. d. 6 1/2 inches
3. e. 6 5/8 inches
3. f. 6 3/4 inches
3. g. 6 7/8 inches

Unit 26

Reading a Leveling Rod Having 1/8-Inch Graduations

1. 5/8 inch
3. corner #4

Section 7
Computing Geometric Measure, Area, Volume, Mass, and Force

Unit 27

Area of Rectangles, Triangles, and Circles

1. 8 square feet
3. 28.26 square feet
5. 1,750 pounds
7. 23.766 square inches
9. 27.641 square inches
11. 11%
13. 1,600 square feet
15. 200 linear feet
17. A minimum of 20 inches in front of and 12 inches beyond each side
19. 50.24 square inches

Unit 28

Volumes of Cubes, Rectangular Prisms, and Cylinders

1. 512 cubic inches
3. 1,536 cubic inches
5. 1385.01 cubic inches
7. 0.78 cubic yard

Unit 29

Weight (Mass) Measure

1. 829.40 pounds
3. 37,120 pounds
5. 358.35 pounds
7. 8.32 pounds

Unit 30

Force Measure

1. 81 foot-pounds
3. 370 foot-pounds
5. 270 foot-pounds

Section 8
Formulas to Align Masonry Walls

Unit 31

Square Columns and Piers

1. Diagonals measure 21.63 inches
3. Diagonals measure 6.36 feet
5. Diagonals should measure 89.6 inches